*talking*SCIENCE

*talking*SCIENCE

Edited by Adam Hart-Davis

John Wiley & Sons, Ltd

Published in 2004 by John Wiley & Sons, Ltd, The Atrium, Southern Gate
Chichester, West Sussex, PO19 8SQ, England
Phone (+44) 1243 779777

E-mail (for orders and customer service enquires): cs-books@wiley.co.uk
Visit our Home Page on www.wiley.co.uk or www.wiley.com

Other Wiley Editorial Offices

John Wiley & Sons, Inc. 111 River Street, Hoboken, NJ 07030, USA

Jossey-Bass, 989 Market Street, San Francisco, CA 94103-1741, USA

Wiley-VCH Verlag GmbH, Pappellaee 3, D-69469 Weinheim, Germany

John Wiley & Sons Australia, Ltd, 33 Park Road, Milton, Queensland, 4064, Australia

John Wiley & Sons (Asia) Pte Ltd, 2 Clementi Loop #02-01, Jin Xing Distripark, Singapore 129809

John Wiley & Sons Canada Ltd, 22 Worcester Road, Etobicoke, Ontario, Canada, M9W 1L1

Wiley also publishes its books in a variety of electronic formats. Some content that appears
in print may not be available in electronic books.

Library of Congress Cataloging-in-Publication Data

A catalog record for this book is available from the US Library of Congress

British Library Cataloguing in Publication Data

A catalogue record for this book is available from the British Library

ISBN 0470093021

Typeset in 11/15 pt ITC New Baskerville by Sparks, Oxford – www.sparks.co.uk
Printed and bound in Great Britain by TJ International Ltd, Padstow, Cornwall
This book is printed on acid-free paper responsibly manufactured from sustainable forestry
in which at least two trees are planted for each one used for paper production.

10 9 8 7 6 5 4 3 2 1

Acknowledgements

First, I would like to thank Rainbow Media, for organizing the 'one-on-one' television interviews in the MagRack series *Maximum Science,* from which this book emerged, and the Rainbow Media staff, who arranged for transcripts of the interviews. Next, I thank all the scientists who came along and talked to me; I was highly privileged to be given so much time by such busy and interesting people. I thank Emily Troscianko, who helped me greatly with the initial editing of the transcripts, and Sally Smith, my editor at Wiley, who encouraged and assisted but never demanded. And most of all I thank Tom Levenson, the producer of the interviews, for persuading the victims to agree to appear, and for guiding me with lines of questioning. My only complaint is that Tom would not allow me to ride my bike to the studio in London; he said it was too dangerous, and he wanted me to stay alive ...

Thank you all.

Adam Hart-Davis

Contents

Preface

I have been asked to do a variety of things for television programmes, but few as stimulating and delightful as interviewing a galaxy of scientists for MagRack's series *Maximum Science.* I have always loved science, and I have spent my working life trying to understand and explain its ideas. To have the chance to talk to these stars about their work was an immense privilege and a great pleasure. The interviews were transcribed, and tidied up, and this book is the result.

Here are top-rank scientists talking informally about subjects ranging from deep ocean trenches to deep space-time, and from illogical sex to computers with attitude. They speak with humour, with passion, and with deep understanding. My favourite assertion came from the Astronomer Royal, Sir Martin Rees: 'Cosmologists are often in error but never in doubt.' Here are a few other gems.

Lord May, President of the Royal Society: 'I went to university to become an engineer, and while I was at university I discovered there was this wonderful world where you can spend your life as a researcher. I always liked playing games; I've played chess, I've played bridge, and I played snooker a lot when I was at university. But I discovered there was this hedonistic life where someone was willing to pay me to spend my life playing games with nature, where the name of the game was to try to work out what the rules are.'

John Maynard Smith: 'Sex is a puzzling problem for an evolutionary biologist. Why do we bother? The orthodox answer is that a species that has sex can evolve much faster to meet changing circumstances, because genes that arise in different ancestors by mutation can join together in a single descendant, whereas if there were no sex, there would be no way these good genes from different lineages could ever get together.'

Richard Gregory: 'Science is much more interesting than Harry Potter – than magic. A test tube and a microscope and telescope – they're much more powerful than magic wands, much more exciting in what they can do.'

Jocelyn Bell Burnell: 'People think you shout "Eureka!" and run naked through the streets! But it wasn't like that. This was a worrying period because this signal was *so* unusual, *so* bizarre, that we nicknamed it Little Green Men, because if it's not Earth men and women, maybe it's another civilisation out there!'

Sir Michael Berry: 'Quantum mechanics has democratized music. You can go anywhere in the world, in the jungle, in the mountains, in Antarctica, and listen to almost perfectly reproduced music.'

Rosalind Picard: 'If you're wearing your computer it can watch you and say, "Oh dear, he's looking like he's getting agitated." Maybe the computer has noticed that in Word you get annoyed every time you see that little paper clip, the office assistant, and it could offer to turn it off for you – completely disable it so you never have to look at it again ...'

Richard Dawkins: 'I feel I have a mission to persuade my scientific colleagues to write their science as if they had a lay person looking over their shoulder, not to write in a language which is completely opaque to other people. I believe they'll do better science if they do that.'

Loren Graham: 'Palchinsky was a patriotic Russian; he shared the desire of the Soviet government to industrialize and become a great industrial power. He just wanted to make sure, when they went about such enormously costly activities as building the world's largest steel mill, or building the world's largest hydroelectric dam, that they did it the right way. They arrested him. They accused him of trying to overthrow Communism and re-establish capitalism in Russia, and they shot him.'

Eugenie Scott: 'My colleagues and I are trying very hard to keep evolution in the public schools. Evolution is a basic foundational idea of all sciences. The intelligent-design people say that some biological structures cannot be explained through natural process, and therefore have to be explained by "an intelligence". They're not talking about little green men. The intelligence they're talking about spells his name with three letters, and the first one's a capital G.'

Lewis Wolpert: 'I've been offering a bottle of champagne to anybody who will tell me one new ethical issue that cloning a human being would raise. I also say if you can't think of one, give me two bottles. I'm against cloning because of the dangers of abnormality. You see, in order to get Dolly, they had to do about 270 trials. From what we know about the way the embryo develops, using the method of cloning to get a child, I think there's a very high risk of things going wrong and having abnormalities.'

Eric Lander: 'The public human genome project, of which I'm a proud member, felt strongly that this information belonged freely in the public domain so that anybody could use it. We put it on the Internet. Every 24 hours the team posted all the new information on the Internet. Tens of thousands of hits on these Websites occur every day. This is the greatest time to be in the game. I can't imagine a better time for a young person to want to go into science.'

Anyway, here are the scientists, talking about science in the raw. I hope you enjoy their ideas as much as I have.

Adam Hart-Davis, Bristol, May 2004

chapter **ONE**

Jocelyn Bell Burnell

Discovery of pulsars

Susan Jocelyn Bell (Burnell) was born in Belfast, Northern Ireland, on 15 July 1943. Her early interest in astronomy was encouraged by the observatory staff at the nearby Armagh Observatory, of which her father was the architect. She graduated in physics at the University of Glasgow and did a PhD in radio astronomy at Cambridge University, constructing the radio telescope with which she would discover the first pulsar, a star that releases regular bursts of radio waves. The pulsars appeared only as an appendix to Bell Burnell's PhD thesis, but opened up a new branch of astrophysics; her supervisor was awarded the Nobel Prize in 1974. She worked as a scientist at the Mullard Space Science Laboratory of University College London, as a senior research fellow at the Royal Observatory in Edinburgh, Scotland, and chaired the Physics Department at the Open University in the United Kingdom. Currently she is Dean of Science at the University of Bath. For a number of years she helped to organize the Edinburgh International Science Festival, being particularly concerned with the public understanding of physics and astronomy.

ah-d

Jocelyn, you discovered pulsars, which were an entirely new sort of object in the universe; do tell me about it! You were, what, 25?

jbb

Yes, I was in my early twenties, working at Cambridge University for a research degree, a doctorate, and writing my thesis. I'd spent a couple of years as one of a group building the equipment – a radio telescope.

ah-d

One of those dish things?

jbb

Well, ours wasn't. Ours looked more like a hop-field or something like that; thousands of wooden posts with wires strung along between the tops of the posts. But we did build it ourselves.

ah-d

Literally?

jbb

Literally, yes. I became very good at swinging a sledgehammer, which was not one of the skills I expected to get as a research student! But useful nonetheless.

We spent two years building this radio telescope, then the rest of the crew moved on to other projects and I was left to run it. Our plan was to survey the sky for some new, exciting objects called quasars – short for 'quasi-stellar radio sources'. They're very, very distinct objects. They're fantastically powerful. They're quite a mystery in themselves even yet. And they'd just been discovered two or three years before I came on the scene. So my project was actually a sky survey to find as many quasars as we could find from Cambridge, position them, and measure their size, that kind of thing. So I had my sights set on the very distant universe.

We were using a new technique; the photographic equivalent is very short exposures. Let me explain. If you took a photograph of traffic signals using a long exposure time you'd get a sort of blur of the red, the amber and the green. But if you want to use a camera to discover how traffic lights work, you'd have to take short exposures.

We were using short exposures to help us pick out these quasars. Unfortunately we picked out something else totally unexpected – a whole lot of locally generated interference: badly suppressed cars, sparking thermostats, all these kind of things!

ah-d

So what do you do about a badly suppressed car?

jbb

Well, basically you hope it goes away! A car probably will, but a sparking thermostat may not. We had to go out with a little radio receiver, with a little aerial like a primitive TV aerial, and scan the horizon listening on the headphones and saying, 'It's that house there!'

ah-d

So you got rid of as much local interference as you could, but how did you listen for your quasars?

jbb

Well, you don't actually listen, because you need a hard copy – something to take away and measure up afterwards. So instead of headphones you normally have the data being recorded. Today it goes straight into the computer. But in those days there were hardly any computers; so it went out on to paper chart – a strip chart that moved under the pen.

ah-d

So you had yards and yards of this paper?

jbb

Miles and miles! We had a hundred feet of chart paper coming out every day, and I ran this survey for six months; so keeping up with the chart paper was one of the nightmares of my life!

ah-d

What was the first thing you spotted?

jbb

When you start working with a new piece of operating equipment you soon get to recognize the various signals it produces – in my case the quasars that I was meant to be studying, and the interference, which is a fact of life for radio astronomers.

But there was occasionally a third kind of signal that didn't look exactly like either of those two. To begin with I just logged it with a question mark: what was it? It was a very small signal; it occupied about a quarter of an inch in this hundred feet of paper that came out. It often wasn't there at all, and when it did appear it was a very weak signal, close to the level of what I could detect. So it was intermittent and 'scruffy'.

When you're a trained scientist, your brain stores problems. And I had this lodged in the back of my brain. After seeing this quarter-inch of scruffy signal a few times, my brain said, 'You've seen this before. You've seen this before from that bit of sky, haven't you?' Then recovery was quite easy, because I'd filed the rolls of chart paper by the bit of sky that I was looking at.

So I went back to the old charts of that bit of sky. And it *was* there, just occasionally. The funny thing was, it was coming up in the middle of the night!

ah-d
Well, don't they always – in astronomy that is?

jbb
No! I deliberately went into *radio* astronomy because you do it in the daytime, Adam! I like my bed! You can actually do radio astronomy by day and night. The sun doesn't ruin the view as it does in optical astronomy.

This was in the middle of the night, and the technique we were using should have produced variable things in the daytime, not the nighttime. So it was a bit of a puzzle. I discussed it with my supervisor, and we agreed that we needed to get the equivalent of a photographic enlargement. In this quarter-inch of scruff the pen was going much too fast, and we couldn't see the detail, because it was all jammed up. We needed to stretch it out by running the paper faster under the pen. However, there was a problem. I was already using a hundred feet of chart paper every day. Clearly we couldn't afford to leave the chart recorder on high speed all day, every day. So that meant I had to go out to the observatory at the right time of night and switch to the high-speed recorder. I did this for a week, two weeks, three weeks, four weeks – and there was nothing!

ah-d
It never came?

jbb
It didn't appear. My supervisor was livid: 'Oh, it's a flare star, it's a one-off thing. It's been and gone and done it, and you've missed it!' After four weeks I got a bit cheesed off, and one day skipped my trip out to the

observatory to go to a lecture in Cambridge. I can still remember that lecture, and it was very interesting. But I was a little embarrassed the following morning when I went out to the observatory, checked the charts of the previous day, and there was a little scruffy signal …

ah-d

No! Life's not fair, is it? Wasn't it Einstein who said, 'Nature is subtle but she doesn't cheat'? It seems to me you were almost being cheated there!

jbb

Well, that's actually typical of research! You have to persevere. I decided I'd better hang around till the right time of day, that day, and then switch over to the high-speed recorder. So I stayed on at the observatory, switched over at the right time, and wondered what would happen. And the pen went 'blip, blip, blip, blip, blip', giving me pulses about every one and a third seconds.

Nothing like this had ever been seen in radio astronomy before. It was most peculiar. As soon as the bleeping stopped I whipped the chart off the chart recorder and laid it on the floor, which was the only place that was long enough and flat enough, and checked the spacing of the pulses. It was indeed very precise.

I managed to get my supervisor on the phone and said, 'Tony, that scruffy signal is a string of pulses one and one-third seconds apart.' Stony silence! 'Oh well,' he said, 'that settles it. It must be man-made. Must be artificial.'

He knew far more astrophysics than I did, which was why he was able instantly, almost intuitively, to say, 'One and a third seconds: there's a limit to how big that object is. It must be no bigger than one and one-third light-seconds across.'

ah-d

Sorry, one and one-third light-seconds across?

jbb

There's a theorem in astronomy which says that if a star changes its brightness, its maximum size is given by how far light can travel in the time it took to change its brightness. So that implies the size of this thing is less than the distance light travels in one and a third seconds. Actually, there's a tighter limit using the duration of the blip. Because the blips were about a quarter of a second long, the object could not be more than one-quarter of a light-second across, which is about 75,000 kilometres, since light travels at about 300,000 kilometres a second.

People think you shout 'Eureka!' and run naked through the streets! But it wasn't like that.

ah-d

But 75,000 kilometres sounds quite big to me …

jbb

Yes, it sounds big, but it's small in astronomical terms.

I didn't see why it had to be man-made, as he said. Anyway, he was interested enough to come out to the observatory next day, at the right time, and stand looking over my shoulder. I realize now, knowing how weak and erratic these things are, how incredibly lucky I was at this point, because it performed again – 'blip, blip, blip, blip', one and a third seconds apart. And this began a very worrying period!

ah-d

Worrying?

jbb

Yes! People think you shout 'Eureka!' and run naked through the streets! But it wasn't like that. This was a worrying period because this signal was *so* unusual, *so* bizarre, *so* unexpected. Nobody had ever seen anything like it before in radio astronomy. And it did look rather artificial. And this one and a third seconds is the kind of frequency you might set on a signal generator to produce pulses.

ah-d

What did you think it was? You must have guessed.

jbb

Our guesses evolved quite rapidly over the next month. We started by suspecting there was something wrong with the equipment. And given that I'd spent the last two years building it, I was really worried! I thought, 'I've literally got some wires crossed. And these bright Cambridge professors are going to discover this, and I'll be thrown out with no PhD!' We persuaded a colleague and his research student to check with their radio telescope and their receiver and, after some scares, they picked up the signal as well. So we knew it wasn't just a fault with my receiver.

ah-d

What about local noise? You mentioned cars and thermostats …

jbb

I worried about that. But then I looked at the exact timing of the signals and I realized that the source was moving around with the stars. The earth rotates in 24 hours, but a particular star returns to the same place in the sky after only 23 hours and 56 minutes. That's called sidereal time, and my scruffy signal was following sidereal time. So it couldn't be Joe driving home from work in a badly suppressed car, unless he was getting off work four minutes earlier each day – and this had been going on for a few months. Unless it was astronomers! Because astronomers keep sidereal time!

We established that it wasn't other astronomers, and then we began to think in terms of something out in space. But we were still hung up on the idea that it was artificial, man-made. And that was when we nicknamed it 'Little Green Men', because if it's not earth men and women, maybe it's another civilization out there! We actually started a test to check up on this. We argued that if it is another civilization, they probably live on a planet which goes round their sun. As their planet goes round their sun, we should see variations in this very precise pulse period – a Doppler shift that would decrease the frequency as they moved away and increase it as they were moving towards us.

And we found a Doppler shift. But it was the motion of the earth around the sun!

ah-d

Wow! So it had to be a stationary object way out there somewhere?

jbb

It didn't seem to be moving, that's right. We also managed to get an estimate of the distance, and it turned out to be about 200 light years away, which puts it way beyond the sun and planets, but still within the Milky Way – within our galaxy – which was much nearer than any quasar.

About four weeks after the first discovery, I went down to Tony's office to ask him about something, and the door was shut, which was a little bit unusual – it was the kind of lab where doors always stood open. I knocked and put my head around the door and he said, 'Ah, Jocelyn, come in – and shut the door.' There was a high-level meeting going on – Tony, my supervisor; the prof, who was the head of the group; and another very senior member of staff. They were discussing how we could publish this result. We didn't *really* think it was Little Green Men; that really was a joke. But we had no natural explanation. What on earth *could* it be? We knew by this stage that not only did it pulse very rapidly and was therefore small, but it also kept a very, very, *very* accurate pulse period – which meant it wasn't showing any signs of flagging, which meant it had vast reserves of energy, which implied it was big in some sense.

ah-d
Sorry, it was small *and* big?

jbb
Yes! It's good, isn't it? So we had to be pedantic scientists and say, 'What exactly do we mean by small or big?' By small we mean small in size, dimension. But when we say it's big we actually mean that it's massive. Not necessarily that it's gigantic in size, but it's got an awful lot of mass.

At the time we couldn't reconcile these two things. We now know that these are incredibly compact types of stars, very, very dense. So they are both small and extremely massive. But at that time we didn't understand this, and we didn't know how to announce this result. We really ought to get on with announcing it. We'd only one of them. It was all curious and troublesome.

Later that night, just a few days before Christmas, came the critical point. I remember that night vividly. The next day I was going back to my family in Ireland, with my fiancé-to-be, and we were going to announce our engagement! So I absolutely had to go!

We had this discussion about publication late in the afternoon, and we didn't resolve it. So I went home to get some supper, *extremely* fed up. Here was I trying to get a PhD out of a novel technique, I'd only six months' money left, time was running out, and some silly lot of Little Green Men had to choose *my* frequency and *my* radio telescope to signal to planet earth!

I got some supper, and then I had to go back into the lab, because every day a hundred feet of chart paper had been pouring out, and I had been spending quite a lot of time on testing this funny signal. So I had fallen miles behind – well at least a mile behind – with the analysis of my wretched chart paper.

So I went back into the laboratory and did some analysis, and at about five to ten that night I suddenly noticed a little bit of signal from another part of the sky which might just be another of these scruffy things. Now, five to ten was a significant time because they locked the laboratory at ten o'clock. And you could be locked in or locked out; the choice was yours, but there was nothing in between! And I knew that this bit of sky, this new bit of sky, would be well positioned to be seen by the telescope at two or three in the morning. So I had to take a very quick decision as to whether I was going out to the observatory at two or three in the morning. I got out all the other charts that covered that new piece of sky, flung them over the floor, and saw: 'Yeah! There and there, and maybe there, very rare, but just occasionally.' And it did look just like that scruffy, unclassifiable signal that was the first one.

So I bundled all the charts up on my desk, and rushed out of the laboratory as the guy turned the key! I went home, got really warmly dressed up, and at 2 a.m. I got on my scooter – very, very cold at two o'clock on a December morning, I can tell you, and it was a six-mile drive out to the observatory! When I got there I groaned, because there was some fault in the telescope which we never found, or I never found, which meant that in cold weather it was only working at about half-power. I knew I wasn't going to find anything new in that mode! So I flicked switches, I breathed on it, and I cursed it! And I got it to work properly for five minutes only, but it was the right five minutes on the right setting! And in came 'blip, blip, blip, blip' – this time one and a *quarter* seconds apart.

ah-d
Ah! So it was definitely different?

jbb
Same family! Same kind of thing! And what was really important, Adam: it could not be Little Green Men! Because there simply could not be two lots of Little Green Men on opposite sides of the sky, both deciding at the same time to signal to a rather inconspicuous planet earth.

It could not be Little Green Men! ... there simply could not be two lots of Little Green Men on opposite sides of the sky, both deciding at the same time to signal to a rather inconspicuous planet earth.

So I went home for Christmas and got engaged and was happy, even though I wasn't allowed to talk about my work that Christmas! Afterwards I went back to Cambridge to find piles of charts on my desk. So I started the chart analysis. And within an hour or two, I found on the one piece of chart two more scruffy signals!

ah-d
I suppose you were sort of tuned to it by then …

jbb
For the first six months we were utterly confused as to what they were. Everybody was. They were clearly small, they were clearly vast reservoirs of energy, so they had to be compact. The most compact type of star known at that stage is a 'white dwarf'. And we wondered if it could be a quiver-

ing white dwarf that was launching shock waves through its atmosphere which somehow were producing radio signals. But that seemed unlikely. Somebody remembered that there'd been a paper about some curious things called 'neutron stars'. But we didn't know what we were dealing with. It was clearly something totally new to the astronomical community, and exactly what had still to be decided. And it was probably a full six months before we finally agreed that it was a previously unseen neutron-star type of thing, only about ten miles across.

ah-d
Ten miles! Oh, it's tiny!

jbb
But it weighs as much as the sun. So that's a thousand million million million million tons!

ah-d
So is that one of those objects that's lost all its electrons and has only atomic nuclei crammed together?

jbb
That's right. The previously most dense things, the white dwarves, still have recognizable atoms, with a nucleus and electrons going round. But if that collapses, the electrons merge into the nucleus, and you end up with basically just nuclear matter. They're probably more accurately described as 'neutron rich' stars, rather than neutron stars, because they're not pure neutrons.

ah-d
But why do they pulse? Why do they send out these radio blips?

jbb
We're not 100 per cent sure of the answer to that. But the picture we have

is that they're a bit like lighthouses. You know, the light spins, sweeping a beam around the sky? This thing spins very regularly, because once you get a star spinning it takes a hell of a lot of effort to change it. It's got a magnetic field which, as on earth, is probably not aligned with the rotation axis – the magnetic pole is offset. And we think the beam comes out of the magnetic poles, north and south. It's a sort of conical beam, like a lighthouse beam, of radio waves.

ah-d

And you found four of them?

jbb

Found the first four. There are 2,000 now.

ah-d

Wow! What an achievement! Tell me, was it tough being a woman? I know it's tough being a graduate student; I've been there. Do you think it was tougher being female?

jbb

Yes, undoubtedly. I was the only female in my Honours Physics class. And every time I entered a lecture theatre all the guys whistled, stamped, catcalled, booed, what have you!

ah-d

What's more, your boss got a Nobel Prize.

jbb

Yes. The Nobel Prize went jointly to my supervisor, Tony Hewish, and the head of the group, Martin Ryle, for two different bits of radio astronomy. That was an interesting and important development, because there's no Nobel Prize in astronomy, and there's no Nobel Prize in mathematics:

because Professor Nobel did not get on with his colleagues, the professors of astronomy and mathematics! So there aren't any awards in those areas. This was the first time the physicists decided that at least some astronomy made acceptable physics, and therefore they could award the physics prize to astronomers.

ah-d

But what about you? You found the things!

jbb

Yes, but I *was* only a student. The fact that I was a woman was less important. I think it was my status as a student. Although I have to say that a female student who gets engaged to be married suddenly gets treated in different ways! At that time there was still this old image of science as being done by grand old men. And under them, these grand old men had a fleet of minions who did what they said and didn't think for themselves.

ah-d

When you got married, did you become a wife rather than an astrophysicist? What difference did marriage make to your career?

jbb

I was in my twenties then and most women older than me did not expect to have careers. They expected to get married, have children, and be at home, be supported by their husbands, and in turn to focus their efforts on supporting their husbands and their children. You put your husband first, the kids second and yourself last. Women younger than me expect to have careers. So my generation represents a turning point. Of my contemporaries at school, probably about half had some kind of career and the other half didn't.

ah-d

But you didn't give up your career; you staggered on?

jbb

I staggered on, yes. It was somewhat of a stagger, particularly once the first child came. Childcare was definitely my responsibility. If the kid was sick, my husband always had a meeting and had to go to work. Trying to juggle childcare and a job was a nightmare! I worked part-time for 18 years, until the kid went off to college. We moved frequently because my husband worked in local government, and the way he gained promotion was to move to another authority. So after five or six years in a place we'd be up and off somewhere else, and once again I'd be writing begging letters to the nearest astronomy or physics place, and frankly getting the kind of job you get when you write begging letters. But if I hadn't had the discovery of pulsars behind me, I wouldn't even have got the jobs, I think.

ah-d

Now, looking back, do you resent all that?

jbb

I am a bit annoyed at one level. I did what I could do. Always life is a compromise, and I found pretty good compromises, given where I was and where society was at that stage. But I was perpetually fighting society's assumptions. You know, 'Your child will be delinquent if you work!' The only condition under which it was acceptable for women to work was if they needed the money. So I would make elliptical statements like, 'Well, I'm afraid I need to work.' And it wasn't actually the money; it was because I was bored –

ah-d

And you loved your neutron stars!

jbb
I enjoyed the science, and I enjoyed the intellectual stimulation. A woman in a suburb of a town, with a baby, without transport, is actually extremely isolated.

ah-d
If you could go back now, to that fateful Christmas, would you do it again?

jbb
I'm not married now, and I'm quite deliberately not married now. That marriage broke up ten or twelve years ago. I find being single, I can be much more selfish about my job. And I'm enjoying that. For the first time I can go after jobs because of what they are, not where they are! I have mobility. But, on the other hand, it's been great, having had a child.

ah-d
So you didn't like being towed around like baggage. You could marry or settle down with a writer or somebody who isn't fixed to a job location; that wouldn't hold you back, except for the childcare problem.

jbb
There's always the childcare problem. Also, we're at a kind of turning point of society, and the pressure is on to find a 'new man': somebody who does not feel that he has to earn the bread, who doesn't feel threatened if you look like earning more than he does, or winning prizes. You've got to have a man who's really comfortable about a woman being successful. I have some doubts just how many men there are like that around.

ah-d
Don't you think that has got better? Don't you think people's attitudes about those sexist things have improved in the last 30 years?

jbb
They have improved, but probably not enough. And I'm now concerned that women who marry and have children actually still don't have a choice, but it's in the opposite sense: they have to go to work because they need the money. Which is equally sad and tough.

ah-d
Now, Jocelyn, you're also a practising Christian, a Quaker. Isn't that unusual? Do you find it sits comfortably with astrophysics?

jbb
It does, yes, surprisingly. I think if I belonged to some other churches it might not sit as easily. But actually being a Quaker is very like being a research scientist in some ways!

ah-d
I don't quite understand that!

jbb
No? Well, it was meant to be a little opaque! As Quakers we are exhorted, each of us individually, to understand about God, what it is God requires of us in this world, God's plans for the world, if I may talk in a rather anthropomorphic way! You have a picture of what you think God is and what the scheme of things is. Then, as you gain more experience, you're expected to revise that picture, adjust it, and go on with that new picture, and then, as you gain even more experience, revise it again.

And this is awfully like the research scientist having a picture of what this curious pulsing star is, and then, as more data comes in, revising the picture, and so on. In both you must not be dogmatic, and you must not have a closed mind. You must be prepared first of all to live with things that you don't understand. You must be prepared to say, 'I don't know, but nonetheless I will go out for lunch.' Or, 'I don't know how the universe started, but I'll keep thinking about it and get on with life.' So you must be prepared to say 'I don't know' and you must be prepared to let go of your lovely favourite picture, the model that you have spent years building up, but which has now been proved to be wrong.

You must be prepared to say 'I don't know' and you must be prepared to let go of your lovely favourite picture, the model that you have spent years building up, but which has now been proved to be wrong.

ah-d

So do you think, then, that God put the pulsars there – for you to find, perhaps?

jbb

I have a distinctly unorthodox view of God and God's job description, if we could use that phrase! I want to stress that this is very much my view, and other Quakers might well be horrified. But because of the astronomical knowledge I have, I have problems with the idea of God as a creator, the prime instigator. I also have problems with the idea that God is in control of the world. That comes not so much through astronomy as just through thinking about the suffering there is. Why doesn't God stop it? If God's a loving God, and God's in control of the world, why are there disasters and people dying and starving?

If scientists find themselves with that kind of conundrum, they go back and check their assumptions – which in this case are 'God is loving', 'God is in control of the world.' I haven't got the guts to abandon a 'loving God' picture; I need that. But I am prepared to try to see what happens if I abandon the idea that God is in control of the world. And I've been working with that hypothesis for quite a while now, and it seems to make sense. So I don't pray to God for good weather for tomorrow's big event or whatever, because I don't think God has that power – or chooses not to use it.

ah-d

So it's a little bit like a *koan*, then – a sort of irrational question to which there is no answer. You're saying you choose to believe in God, but you choose to believe that God is not omnipotent or chooses not to exert the power?

jbb

Yes. Yes, that's right.

ah-d

Does this give you problems in the Church?

jbb

Not in Quakerism, no, because, although we do have corporate beliefs, we also respect each individual's experience. I recognize that I'm probably a bit out on a limb in this, but that's OK.

ah-d

You seem to be out on several limbs, if I may say so! I've read about your discovery of pulsars, but it's absolutely riveting to hear it straight from the horse's mouth, if you'll forgive me for saying that! Could I ask you one last question? When you discovered your scruffy signals you called them LGMs. Did you want them to be aliens?

jbb
No! Definitely not! Just imagine the ballyhoo there would have been!

ah-d
You would have been very famous then!

jbb
But I'd have got no work done!

ah-d
That's true. And you cared more about the work than about being famous?

jbb
My money was running out!

ah-d
But didn't a secret bit of you wish that you'd made contact with another civilization?

jbb
No, I think not. There's so much bound up in that: an awful lot of ethical issues about how you interact with another civilization, for instance. I think I had plenty on my plate, thank you!

ah-d
Do you think there's anyone else out there?

jbb
Oh, the universe is so big there has to be! But whether we can make contact with them is another question. Because of the long, long time for the travel of radio signals and light signals, conversations would be hopelessly slow, to say the least! I'm not sure that we can ever actually talk with them, but there probably are people out there, yes.

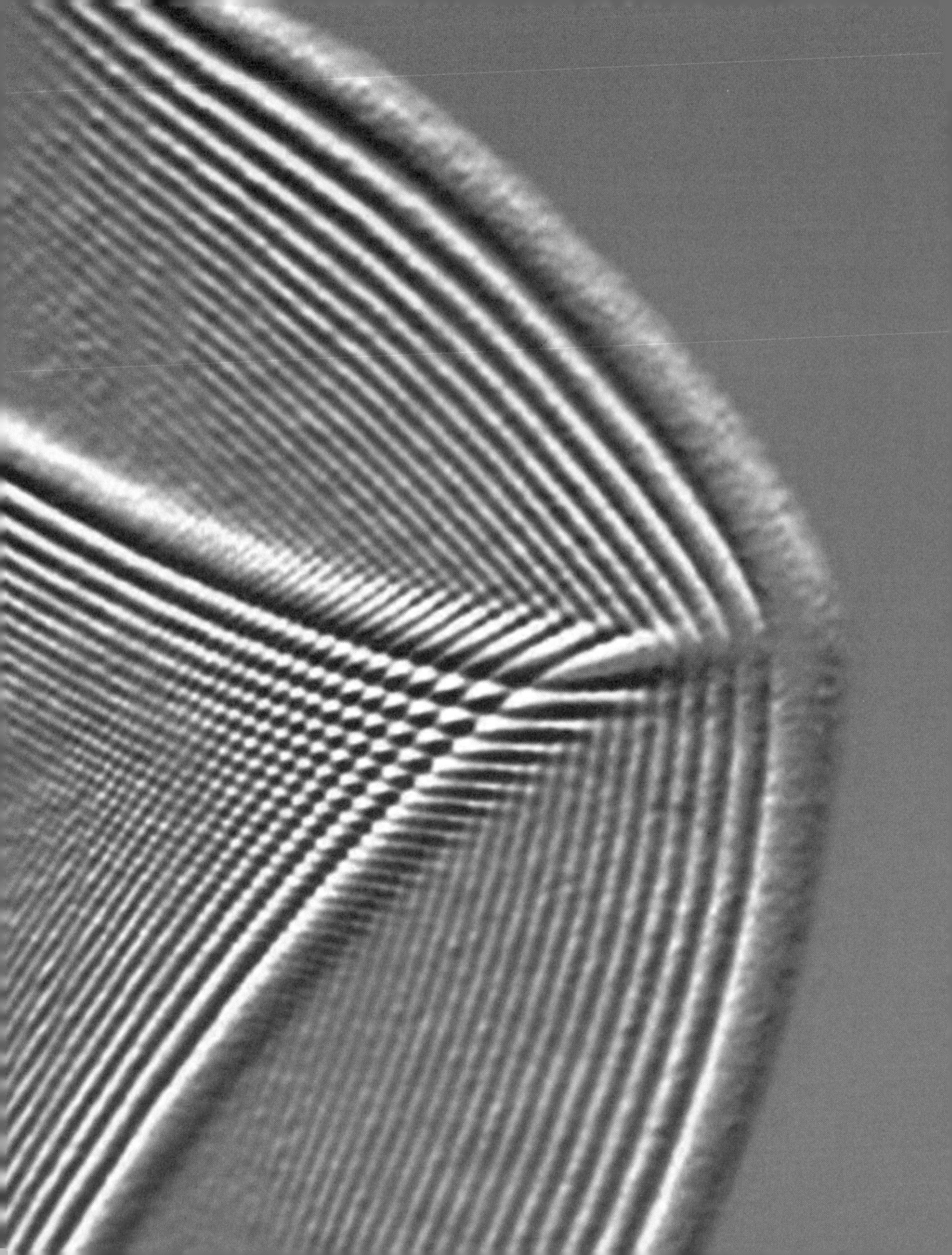

chapter TWO

Sir Michael Berry

Quantum physics

Michael Berry was born on 14 March 1941, in Surrey, England. He left Exeter University in 1962 with an honours degree in physics, after which he did a PhD at St Andrews University and postdoctoral research at Bristol University, both in theoretical physics. Since 1998 he has been Royal Society Research Professor at Bristol University. He works in the borderlands between physical theories – between classical and quantum, between rays and waves – with an emphasis on geometrical aspects of waves and chaos. He delights in seeking out down-to-earth or dramatic examples of abstract mathematical ideas: 'the arcane in the mundane'. When I asked him what he was doing at the moment, he said 'Tying knots in nothing…'

ah-d

Michael, you've been working with light all your life. Didn't Einstein say that he'd been struggling to understand light all his life and failing? Are you following in his footsteps?

mb

No, because what he was referring to was using light to understand fundamental physics, and it's certainly true that all the fundamental advances in physics over the centuries have been accompanied by advances in the understanding of light, by progressing to deeper and deeper levels of theory. But most physicists don't work on fundamental physics, in the sense that they're not seeking the fundamental laws of physics, or changes in them. Once a century you get a change in the fundamental laws of physics, and we're not that lucky; there are too many of us.

So what we mostly do is to spend our time working on consequences of the laws of physics when we know they apply. I'm part of that humble enterprise, but it's astonishing how rich the laws of physics are, and how many unexplored consequences of the laws of physics there are.

That applies to light too, at various levels: at the level of *ray optics* where you just think of the light travelling in straight lines, as through your telescopes and binoculars; at the level of *wave optics* where you think of interference of light and diffraction, or wave properties; and at the level of *quantum light* where you think of the individual photons and the quantum randomness. The theories are so rich, the mathematical formalisms are so dense, that to get out what they contain is an ongoing activity.

ah-d

Do you go into the lab every day and do experiments?

mb

I'm a theorist. I do have a lab in my house where I play. I do what you would call kitchen table experiments, and I'm lucky; there are still new things you can do on the kitchen table. But really I'm a theorist. I sit and think about light.

ah-d

You say it's all about waves, but also about quantum mechanics? Isn't there a discontinuity there, that quantum mechanics doesn't always come in waves, because there are jumps and things?

mb

That's right. It's one of the beautiful disjunctions in physics that you have waves and you have particles. A lot of my research is about what's called the classical limit of quantum mechanics, or the semi-classical limit of quantum mechanics: understanding how it can be that although the deep level is quantum, you don't use the Schrödinger equation to calculate the path of a spacecraft or a satellite, or indeed to ride a bicycle, although these are quantum objects. To understand how it is that somehow, although the theory and the description and the formalisms look so different, it's sufficient to use classical mechanics when things are big and heavy – this is part of my life's work. And it's complicated. The limits aren't straightforward – and that's what is so beautiful about them.

ah-d

Does quantum mechanics ever impinge on our lives?

mb

Oh yes. Let me give you one example: the compact disc player. That has transistors in it, and transistors were a deliberate and direct application of the quantum mechanics of electrons in periodic structures – crystals – to technology. The CD player also has a laser in it, and the laser was a direct application of Einstein's 1918 quantum theory of light.

Now, think what the CD player does. You can go anywhere in the world, and listen to almost perfectly reproduced music. You can be in the jungle, in the mountains, in Antarctica. Quantum mechanics has democratized music. That sums up what quantum mechanics can do.

ah-d

All right, I suppose there's a little bit of quantum mechanics in our lives. But there isn't any relativity in our lives, is there?

mb

Oh yes, not much, but there is. In your car, or soon to be in your car if you don't have one already, is a GPS receiver. The Global Positioning System tells you anywhere in the world where you are within five to ten metres' accuracy. That works because the signals are beamed down from satellites, and inside the machine geometric calculations are done, to work out where you are. Those calculations have to incorporate corrections for the propagation of light that come from Einstein's relativity – about ten different corrections have to be incorporated.

I spoke to the physicist who advised the NASA engineers in the 1970s about this system and he said they were very resistant to the idea that they had to include the relativistic corrections. But if they didn't, you'd be kilometres in error by the end of the day.

So your GPS receiver is a relativity machine. You hold in the palm of your hand the first consumer device that incorporates a deliberate application of Einstein's relativity.

ah-d

The last time I saw you on the web you were levitating frogs. This doesn't appear to be anything to do with physics or waves.

mb

No, it isn't. It's mechanics – beautiful mechanics. There is a toy called the Levitron, which is a commercially available spinning top that levitates, floats in air above a magnetized base. It's astonishing to see this thing floating. When I bought one I had to understand how it worked; my immediate reaction in the shop was, this is a large-scale analogue, a macroscopic, human-sized analogue of the tiny traps that atomic physicists now use to hold particles in place – electrons, neutrons, atoms – while they examine their properties.

You can hold an electron in a trap for weeks and study its properties. You can see a single atom by shining lasers at it, because it continues absorbing and emitting light. You see this gleaming speck. I have seen that, and when I saw the toy I thought, here is a macroscopic analogue.

The shop where I found it is the best scientific toyshop in the world. It's in Zurich, just below the apartment where Lenin lived when he plotted the revolution. I saw this thing and I had to have it, and I said to the assistant, 'You know, this looks like a macroscopic analogue of all these traps.' She gave the most disappointing reply, absolutely squashing all my enthusiasm: 'Oh, we don't want to know about those technical details; here, we like to understand the universe as a whole.'

Anyway, I had to understand this thing, but even the people who invented it – and there's a controversy about who did actually invent it – didn't really understand the dynamics on which it was based. And they were very clever people – this is not a criticism of them; they tried where physicists would not try. I worked on the theory of it and I'll tell you in a minute what the significant point was.

I gave a lecture somewhere, and mentioned this toy, and somebody said, 'Do you know, there's this guy in Holland, Andrey Geim, who's levitating frogs?' And I said, 'No, that's impossible', but I contacted him, and found he had indeed levitated frogs, but without fully understanding the theory, and so we worked together and wrote a paper. So it's his experiment actually; I just thought about it.

To me, this is what is really exciting about physics. It's the unexpected connections you get between completely different ideas.

The point is this. A frog is mainly water, as we are. Water, like almost everything, is very weakly magnetic. You think of it as non-magnetic, but actually when you bring a magnet close to almost every substance, it induces a tiny magnetism opposed to the one that you excite it with, and therefore it repels.

That repulsion is tiny; you normally don't notice it. But it's been known for a long time, more than a hundred years, that if you had a strong enough magnet you could counter-balance gravity. That's straightforward. Similarly with the spinning top: that's magnetized, and it repels.

The problem is how it can be stable. There's a theorem, called Earnshaw's theorem, which says – and it's been known since 1842 – that no combination of electric, magnetic, and gravitational forces alone can hold a set of objects together stably in space. The forces can balance, but the slightest change upsets the balance. It's like a pencil balanced on its point; the forces balance, but the pencil doesn't stay up. Likewise, if you try to balance one magnet above another, it will turn over and fall down.

So the spinning-top people thought, can we spin the top, so that gyroscopic effects prevent the top from turning over? That will keep it in the right orientation to be held stably. When they tried, they found it was very hard; it took about a year of experimenting. The problem is that this spinning ensures that it's orientationally stable, but not that it's positionally stable. It won't fall off, but it can still slide off – and it almost always does.

So this Levitron toy is very hard to operate, but they did find a tiny little pocket of stability. Likewise with the magnetic frog. The spinning top is actually little tiny electrons moving around in the atoms. That's how it works.

So Andrey Geim succeeded, with great difficulty. He had to tune the magnetic field. He used a big Perspex tube, and an immense amount of electricity to generate the magnetic field. He found he could levitate water droplets – beautiful little spheres – hazelnuts, rose thorns. Then he thought, why not something alive? So he put in a very small frog, about a centimetre across, a baby. It seemed to suffer no ill effects. It hopped away afterwards quite happily.

ah-d

But wasn't it cooked? If you're inducing a lot of currents in the water, isn't it like putting it in the microwave?

mb

No, because they're not conduction currents. They're just the summation of all the little circulating currents in all the atoms, which are there all the time anyway. So they don't rub against each other and cause friction as a macroscopic, continuously flowing current would around the outside. It's not a current that excites resistance and therefore heat.

Earnshaw's theorem was important in the history of science because when, in the nineteenth century, people first began to think about how matter might really be made of atoms, the only forces they knew about that would hold atoms together were gravity, electrostatics and magnetism. So they naturally thought some combination of these forces would be involved.

Earnshaw's theorem proved that this couldn't be the case, that there had to be other, more complicated forces, and also that these things must be in motion. We now know this is true. People such as Lord Kelvin, and James Clerk Maxwell were greatly influenced by this little theorem, which comes up again in levitation.

ah-d
Have you done any levitating yourself?

mb
When you stand on the ground, you're levitating, but you're nano-levitating, because what holds you up are the forces between the ground atoms and the atoms of the soles of your shoes. They have a scale of a few nanometers, but still, it is empty space between them. So, standing is nano-levitating, and there's no energy needed to do it.

ah-d
You say that when you stand on the ground you're not touching it?
Why don't we fall through the floor?

mb
Because matter in bulk is impenetrable. This isn't obvious, because we know that matter is mostly empty space. The nucleus is ten thousand times smaller in linear dimensions than an atom, and the electron has no size yet determined. The reason why matter is impenetrable and why atoms don't collapse down to nothing, and why all the electrons and big atoms don't collapse into the lowest orbit of quantum mechanics and

make the atoms much smaller, is something called the Pauli exclusion principle.

This is an assertion that no two electrons can exist in the same quantum state. This applies because electrons are absolutely identical particles. There are curious restrictions on the motions of identical particles in quantum mechanics. When you apply relativity and quantum mechanics together, you can make arguments as to why the Pauli principle must be true. But there is actually a much more beautiful explanation, still controversial but almost certainly having something to do with the truth, which connects this strange principle – that makes matter impenetrable to astonishing geometrical properties in ordinary space – to properties of twists and turns.

This strange geometry, which in its quantum manifestation is responsible for the exclusion principle of identical particles, has to do with turns in space. You might think that turning objects in space is totally understood, ancient geometry. Put it this way: I have something on my desk – a glass of water for instance – and I say I'm going to go out of the room and then come back again, and I want you to ask me whether you've rotated it by one complete turn or not.

And when I come back I can't tell. I can't tell as long as the object is not connected to anything. But if it is connected to something – let's say the buckle of my belt – of course I can tell if you've turned it once because I can see a twist in the belt. It's the residue of the turn. If I turn it again there's another twist, so you might say that by counting the number of twists I can tell how many times you've turned it.

The astonishing thing is that this is true, provided I'm not allowed to move the object as well as turning it. If I'm allowed to move it then I can tell whether its been turned once or three times or five times or seven, an odd number but not an even number, because there's a strange arithmetic for turns of tethered objects that can be moved in space, which says two equals zero. Aren't we physicists crazy?

For turns in space of tethered objects, two equals zero. Now, it turns out that in quantum physics there are two kinds of particles. There are ones that behave as though they're tethered – electrons fall into this category – so that you can tell when they've been turned once, but not twice. And there are ones that behave like ordinary, untethered objects – photons, for example, and alpha particles – where you can't determine any number of turns.

With the electrons, you might ask what these turns have to do with indistinguishability, and that's another little trick. Suppose I've now got two of them, with imaginary string between, and I exchange them, now I've made a twist. So exchanging two particles is like putting a twist in the string. To exchange two of them is like turning one of them. The point is that if I want to ask what a quantum system can do where you can tell the difference after one operation – after one turn but not after two – the essential fact is that quantum systems describe by waves. A wave is a number reflecting the condition, and the only thing you can do to a number that's the same after you do it twice but not after you do it once is change its sign, multiply it by minus one.

This is the mathematics of Pauli. If you exchange two objects, you multiply their wave by minus one. Now, suppose they're in the same place, so that the exchanging of them does nothing. The only quantity that is unchanged by multiplying by minus one is zero. So there's zero probability of finding it there. Pauli, explained in a flash. To me, this is what is really exciting about physics. It's the unexpected connections you get between completely different ideas. Who would think that the impenetrability of matter could be connected with silly conjuring tricks you can do with belts? I think it's wonderful.

ah-d
How did you get into this? How did you start being interested in science?

mb
That's a hard question. I was interested as quite a young lad, seven or eight, in astronomy. Like a lot of kids, I was turned on to science through astronomy, and that sort of led into physics when I got to university. But astronomy was really the main interest; I was fascinated by the fact that you could learn something about these vastly distant realms, that you could know the composition of the stars and understand their orbits and so on.

And it became clear to me that this is a mathematically based understanding, and that therefore I had to learn a bit of mathematics, and so I did. Then, actually quite a bit later, I realized that my abilities, such as they are, lay in the mathematics. I never officially learned mathematics, apart from the very elementary courses that you get as an undergraduate. I picked it up myself, and now that's what I do. My work is a curious interplay of mathematics and pictures of reality.

It could be that as a child, when my parents were fighting all the time and eventually got divorced, I thought of astronomy as something pure and remote and untouched by all of these horrible concerns and things that worried me daily, domestically, and that I was therefore attracted to this realm. One would need to be more of a psychologist than I am to know whether that's true or not, but I've wondered about it. But if that was the original motivation, it certainly didn't survive, because science is quite different for me today. I regard science now as earthy, strongly interactive, immediate, hands-on, connecting with other scientists with whom one endlessly talks and converses. Very, very social.

ah-d

Your original plan was to escape from reality. What sort of household was it, apart from being uncomfortable?

mb

We weren't rich. We weren't very poor, but it was a working-class Jewish family in London; my parents had grown up in the East End of London and had moved out. My father was a taxi driver. My mother was a dress-maker. There was only one other person in the family who was educated; a cousin of mine, a Professor of English in Indiana, now retired. He was very good to me. He opened my mind to the fact that there's a world of culture. But he wasn't a scientist. The science came from me, but I can only speculate how. Of course I benefited greatly from being in a time and place where we had free education. We could be born into any background and get educated at school and at university without having to pay for it, which my parents never could have done.

ah-d

A few years ago you got knighted. Not many physicists pick up knighthoods, do they? What did you have to do to get knighted?

mb

I wish I knew. Usually with physicists it's because they've served on some government committee, which I've never done. I got a nice letter from the Prime Minister – that was John Major – which just said 'physics'; so your guess is as good as mine.

ah-d

I'd like to go right back to the thing I used to talk to you about many years ago, which is the swirling light patterns on the bottom of swimming pools.

mb

One of the areas that I'm studying these days is the connection between the optics of rays and the optics of waves. Mathematically it's very closely analogous to the connection between classical mechanics and quantum mechanics. In classical mechanics you've got trajectories and in quantum mechanics you've got quantum waves. In geometrical optics you've got rays that are like trajectories in lines along which you imagine the energies travelling. And in wave optics you've got light waves: electromagnetic waves. These connections are very similar and surprisingly rich. In order to understand these connections we have to understand what are called singularities.

The most interesting things about the theories are the places where they predict something is infinite. But then the theory is wrong and has to be replaced by something deeper.

Singularities are where some physical theory predicts infinity. I'm not religious, but I could say God has a love-hate relationship with singularities. What I mean is that she gives us these theories that we somehow stumblingly discover and the most interesting things about the theories are the places where they predict something is infinite. But then the theory is wrong and has to be replaced by something deeper.

In light, a good example of singularity is a rainbow. A rainbow is, as Francis Bacon said, the play of sunshine on a dripping cloud. It's light rays bouncing around inside water droplets and coming out again. Descartes first understood this. He understood that although the drop is uniformly illuminated by the sun, the light that comes out is focused in direction.

Focusing is a singularity. It means that if ray optics were exactly true, you would look in a particular direction through a coloured filter and see an infinitely bright band of light for each colour. Of course you don't – it's smoothed out – and one of the reasons it's smoothed out is because of wave effects. Sometimes you can see these wave effects directly inside the main rainbow (not the secondary, which is some way away); just in the main rainbow you can see one or more little alternations of each colour. The colours don't seem to be in quite the right order, as you learned at school: red, orange, yellow, green, etc. That's because you're seeing an interference effect of waves associated with rays, two different rays that come out in the same direction from different parts of the drop.

The laws of physics have been known in ray optics for 300 years, and in wave optics for 200 years, but explaining singularities needed some new mathematics, called catastrophe theory. It was only in the 1970s that the full mathematical understanding was attained of how the laws of refraction can give rise to focusing when you don't have any symmetry. This is what is called natural focusing.

The flickering pattern of light on the bottom of a swimming pool is a good example. The pattern is dominated by lines of focused light, called caustics. An even better example is bathroom window glass. It's knobbly, and the knobbles have a scale of about a millimetre. If you shine a laser pointer through the bathroom glass what happens is that each ray in the beam passes through a different part of the landscape of the glass. And on the wall, in a darkened room, you can see places where more than one ray hits, and in such places you get interference, as a system of fine lines. The boundaries, which are the very bright lines, are caustics. These are the singularities of optics. As you move the laser beam across the glass you illuminate different bits of this random landscape and make intriguing moving patterns on the wall.

Isaac Newton would have died for one of those little laser pointers; it's the closest you can get to a pure perfect ray of light, one of the rays that he imagined – but it isn't one ray, it's a bundle, about a millimetre across, of infinitely many rays.

The new mathematics of catastrophe theory enables us to describe the structural elements of these wonderful patterns. There are little pointy things called cusps. They keep coming and going, but you notice that near them the wave pattern is always the same – one of the standard patterns predicted by the mathematics. You can produce them on the computer and check that you've got it right by comparing them with reality.

If you take one of the smooth curves and look at the interference lines that decorate it, these are the patterns that decorate rainbows. You normally cannot see them very clearly in rainbows, because the different colours get mixed up and the water drops have different sizes and the sun isn't a point, it's a disc – effects that blur the rainbow interference. But you can still see it sometimes if you're lucky.

But you don't need a laser to see such patterns. If you wear glasses, and go out at night in the rain, raindrops will fall on your glasses and make little irregular water-drop lenses – lenses on the lenses. If you then look at distant streetlights, you'll see the images of those streetlights drawn out into patterns with the interference included.

ah-d

You seem to think physics is fun. Some people, me for instance, regard it as incredibly complicated and beyond the reach of normal human brains.

mb

When you pursue something, you pursue it and pursue it and of course it gets complicated. Complications are fun. There are some people, I understand, who are interested in sport. I'm not at all, but when I read in newspapers, or I hear on the radio, descriptions of golf or cricket, they seem to be incredibly complicated. People learn these specialist skills. Take a knitting pattern. A knitting pattern is like a computer program. It's a very complicated series of instructions coded according to formal rules which, if you don't understand them, are completely impenetrable. It's a language; you get used to it. That's not to say physics is easy, but it certainly is fun.

chapter THREE

Colleen Cavanaugh

Zoology in deep-sea trenches

Colleen Cavanaugh was born in Detroit. Inspired by her seventh-grade environmental science classes, she studied biology at the University of Michigan and took a course in marine ecology, studying horseshoe crabs at Woods Hole, Massachusetts. In her postgraduate work at Harvard she focused on the partnerships between animals and carbon-fixing bacteria around hydrothermal vents on the ocean floor. In 1992 she won a place on the deep-diving submarine Alvin, and went to the bottom of the Gulf of Mexico off Florida for the first time; she is now Jeffrey Professor of Biology at Harvard. Her research concentrates on the nature and evolution of interactions that allow host and guest to survive in otherwise inhospitable places, studying ribosomal RNA to categorize bacteria and determine their evolutionary symbioses.

ah-d

Professor Colleen Cavanaugh, you get paid for going on ocean cruises and finding bizarre animals. How does a young graduate student get into that sort of life?

cc

Well it all started when the deep-sea hydrothermal vents were discovered. These are hot springs like Yellowstone springs, but at the bottom of the ocean, associated with areas where the sea floor is spreading and new crust is being made. Surprisingly, there are amazing oases of animals with a biomass that rivals that of rain forests. And some of the animals are just incredibly bizarre.

I came into the picture about three years after that. It was my first year of graduate school, and we had a seminar called 'Nature and Regulation of Marine Ecosystems'. As part of that, the professors brought in four people who had been working on these hydrothermal vents – an ecologist, a chemist, a microbiologist, and the Curator of Worms at the Smithsonian. This last scientist talked about these giant tube-worms he was working on that can grow up to two metres long.

ah-d

Two-metre-long worms?

cc

Yes. Well, at least the tubes. The worms may actually be smaller within them. He gave essentially a micron-by-micron lecture through this tube-worm. And fortunately I was still awake when he showed a slide of a cut right through the middle of the worm.

These tube-worms were unusual. The scientists had established that the food chain was being supported by chemosynthetic bacteria – bacteria living on pure chemistry. But these tube-worms were completely bizarre in that not only are they huge, but they also completely lack mouth and gut.

ah-d
So they can't eat?

cc
Well, they had cousins that were the size of a piece of hair and had been shown to be able to take up organic molecules across their skin. So that's what the scientists assumed they were doing. However, the first measurements of organic carbon dissolved in the seawater at the vents were very low. So the question was, how were they making a living?

During the course of this lecture he showed a slide of what is called the trophosome. It's this brown spongy tissue that fills the bulk of the coelomic cavity of the animal. And it was named the trophosome from studies on an earlier worm, because the gonads, the eggs, and the sperm were also embedded in this tissue.

So it seemed as though this tissue was feeding the eggs and sperm. But when he was dissecting them, he discovered a lot of white crystals in the tissue, which turned out to be pure, elemental sulphur. This is where I jumped up and said, 'Whoa. It's perfectly clear. There must be sulphur-oxidizing bacteria that can use the hydrogen sulphide that's in the water from the vents, and react it with oxygen, and get energy from that reaction to fix carbon dioxide.' In other words, there must be symbiotic bacteria within the tissue that are feeding the animals. They would be similar to symbionts, e.g., algae – the photosynthetic organisms that live in coral and feed the animal internally. And he effectively said, 'No, no. Sit down, kid. We think it's a detoxifying organ. Hydrogen sulphide is a potent toxin. It binds to your haemoglobin. It binds to your cytochromes and renders them unusable, so you can die. In fact, hydrogen sulphide is as toxic as cyanide.'

He continued, 'We think it's a detoxifying organ, and it's oxidizing sulphide to elemental sulphur, which is non-toxic.' And I said, 'Well, that's fine. But if you have bacteria doing the sulphide oxidation, they can detoxify the sulphide, make energy and fix carbon dioxide, and feed this mouthless, gutless animal.' Eventually I was able to convince him, and he sent me a small piece of tissue to examine for the presence of bacteria.

I was only a first-year graduate student, and I had not even had a microbiology course, but I had done a lot of work on free-living bacteria in the environment. So I used various forms of microscopy looking at this tissue, and couldn't figure it out, because it looked like bunches of grapes, not at all like bacteria. Holger Jannasch, a vent microbiologist, reminded me that when bacteria are in symbiosis, notably living inside of the cell of a host organism, they can lose their cell wall and become quite odd-shaped, because they're osmotically protected by the host cell – that is, they are held together by the body they are living in, and don't need their own 'skin'.

I ran all sorts of tests – I used nucleic acid stains to show there was DNA or RNA in the cells, I ran a lipo-polysaccharide test, I learned transmission electron microscopy – and I kept accumulating evidence.

And then someone else found an enzyme that is only known to occur in organisms that are autotrophic – that is, self-feeding. They can take carbon dioxide out of water or the air and fix it into organic molecules. So to me finding it in the animal tissue was a clincher. And together we wrote a series of papers establishing that there was this symbiosis between sulphur-oxidizing, chemo-autotrophic bacteria and these giant tube-worms.

ah-d

Amazing. Do you always get everything right when you jump in like that?

cc

No, no. In fact my mother said 'How did you walk in under all these professors' noses and discover this?' And I said, 'Well, I think it's because my background was in ecology. And so I wasn't a zoologist just looking at the animal or a microbiologist just looking at bacteria but rather, had a more "systems" view.'

I wasn't a zoologist just looking at the animal or a microbiologist just looking at bacteria but rather, had a more 'systems' view.

ah-d

Now, please can you take me on a research trip? We're talking the bottom of the deep ocean. The middle Atlantic, say, where the ocean floor is actually spreading, and there is hot stuff coming up from underneath. How big are the vents?

cc

Well, the vents are in fields. So they can be a very small … just a few metres to tens of metres. And they're found along the mid ocean ridge system, which runs all the way around the earth along the ocean floor.

ah-d

Have you actually been down to the vents? How do you get there? Now, when you get down there, can you actually see stuff coming up?

cc

Yes, I have dived to the vents. We use Alvin, the little submarine run out of Woods Hole. You get in the sub while it is on the deck of the ship, and then it is launched using a big A-frame over the stern. Alvin sinks of its own accord, and is not attached to the ship. It actually descends like a gyroscope, which I didn't realize until I saw videotapes of our descent from outside. I didn't appreciate that because when you're inside you don't know. As you go down, it gets darker and darker until it is pitch black, and then you begin to see bioluminescent organisms. You see them all the way down – and we're typically working at about 2.7 kilometres' depth. The deepest site that I have dived to is 3.7 kilometres; that's over 2 miles.

ah-d

Wow. So the pressure is enormous down there.

cc

Yes. But the sub is a sealed titanium sphere, and the pressure inside stays at one atmosphere.

ah-d

How big is it? How many people inside?

cc

There are three people. There's a pilot who sits in the centre. And then the two observers sit on the two sides, port and starboard, and you are looking down at an angle through the three portholes. There are video cameras mounted on the outside of the sub, and you watch the video screens as well.

ah-d

And presumably remote arms so that you can get samples and so on.

cc

Yes. Alvin has manipulators that the pilot uses to collect samples. And the sub has a large basket that contains the sampling equipment on the front.

ah-d

What do you see when you get down there? You have searchlights, presumably?

cc

Yes. Alvin has very bright lights. To locate the vents what the pilots and observers look for when they get close to the bottom are the animals, which are found in greatest numbers right at the vents. Prior to the discovery of the vents, life in the deep sea was thought to be very sparce, given that it is dark, cold (~2°C – just above freezing) and under very high pressure. Deep sea life was known to be very diverse, but low in biomass: in large part due to the general lack of food, since primary production by algae is limited to the surface waters where there is sunlight. By the time it rains down it is decomposed and there's very little left.

So when the geologists and geochemists went went looking for these vents, based on their theories of sea-floor spreading, they were totally astonished to find these dense clumps of tube-worms that were just beautiful. Indeed, the scientists gave these vents field names like Rose Garden and Garden of Eden.

ah-d

These tube-worms, what do they look like when you see them?

cc

They look like long, very white tubes with red tips. The tubes are made of a chitinous-like material, like your fingernails, flexible when they're alive, but they become hard when dried.

ah-d

Like plumbing pipe then?

cc

Yes. The worms have these beautiful red plumes that they can move in and out of their tubes, which are effectively their gills or their lungs. If you look closely, they look like little fans because they have individual filaments, but from a distance they look almost like flowers. The tube-worms have an amazing blood system and the red colour is due to their high haemoglobin content. So there's tremendous gas exchange. They take up oxygen and hydrogen sulphide from the mixing deep-seawater and vents fluids respectively.

ah-d

Where does the oxygen come from?

cc

People typically think there's no oxygen in the deep sea. But, in fact, it's highly oxygenated. Water at the poles is cold and becomes highly oxygenated. This water, which is more dense as a result of its cold temperature, then sinks and mixes in the deep sea. And the oxygen stays in solution once it's down there.

The hydrothermal vents, or springs, are associated with areas of mountain chains in the ocean floor where lava is welling up and new crust is being formed. There's a lot of volcanic activity and a lot of fracturing of the rocks. Cold seawater percolates down into the cracks, is charged with volcanic gases, interacts with the basalt at very high temperatures – becoming enriched with many other chemicals – and then finds other fissures where it exits the sea floor. It comes streaming out as 'black smokers', forming chimneys made out of metallic sulphides. When this hot acidic brine hits the cold seawater, various metallic sulphides precipitate out. It looks like smoke, but it's actually particles of sulphides.

ah-d
What sort of metallic elements are there?

cc
Iron, manganese, copper. The black smokers are the very hot vents, at 350 to 400°C. The animals are found where the vents are more diffuse, where there's shallower mixing. So they're approximately at room temperature, which is still very warm compared to the deep sea temperature of near freezing. But it varies moment to moment from 2 up to about 30 to 40°C.

ah-d
Are all the animals in thin lines along the edges of the hole, or are they spread out in fields?

cc
It varies depending on the vent site. At eastern Pacific vents such as those explored off the Galapagos Islands or south of Mexico there are giant tube-worms, giant clams as big as dinner plates, and mussels. The mussels are found in large beds and the clams are often found in crevices. And there are crabs scavenging all over. Then there are numerous other animals, including fish, that colonize the vents.

ah-d
Are there other places in the world where you find these vents?

cc
Yes. They were originally discovered off the Galapagos Islands, and have been explored along other sites in the eastern Pacific, south of Mexico,

up off the coast of Oregon and then in the western Pacific, as well as in the Atlantic. Most recently vents were discovered and explored in the Indian Ocean.

ah-d

Are they all the same?

cc

No, the vents are not the same. The chemistry can vary, and even with similar chemistry, the animals that are found can be quite different. For example, eastern Pacific vents have the characteristic tube-worms, clams and mussels as the obvious macrofauna. And then the Atlantic vents, surprisingly, were dominated by shrimp.

ah-d

But no shrimp in the Pacific?

cc

Well, there *are* shrimp, but not this particular species and not in the same numbers. I mean, in the Atlantic you find spires that would just be totally covered by these shrimp. Spires of sulphides, tens of metres high.

ah-d

Like a sort of cathedral?

cc

Yes. There are actually vent sites that have been named 'cathedral'. On the sulphide spires the shrimp are all jostling for position and actually ingesting these metallic sulphides. In the Indian Ocean, visually the vent sites look very much like the Atlantic, because there are zillions of shrimp. But the mussel species there and other animals are more closely allied with the western Pacific species. So it looks like we have a shrimp species that spans something like 17,000 kilometres across the ocean floor. The ridge segments between the Indian and western Pacific Ocean have not yet been explored, so we can only speculate on the species composition between these sites.

Indeed, this brings up an interesting aspect of deep sea oceanography. We have seen more of the surface of Mars than we have of the ocean floor of our own planet. With the Mars rovers and their cameras, you can see for a long distance because of the availability of light. But seawater attenuates light very rapidly. So you can only see about ten metres distance looking through the Alvin portholes or at the videos taken by Alvin's cameras. So there are huge segments of the Earth's sea floor that we have never seen.

The great advances are often made at the intersections of different fields.

ah-d
What sort of a scientist are you? Are you a microbiologist?

cc
Yes, I would call myself a microbiologist, or more essentially a microbial ecologist, because I'm interested in what bacteria are doing in the real world, their ecology, their evolution, physiology. And while I have focused mainly on bacteria as symbionts, we are now starting to search for those chemosynthetic symbionts in the free-living state.

ah-d
But you seem to get involved in all sorts of things – oceanography, geology, chemistry …

cc

Yes, absolutely right; the joy of ecology is that you are looking at the interaction of organisms with organisms, as well as organisms with the environment, and therefore you really should be a meteorologist, a molecular biologist, a chemist, a geologist. The great advances are often made at the intersections of different fields.

ah-d

So you are interested in how things interact. Tell me about those tubeworms. They depend for their food entirely on the bacteria: how could they have started without the bacteria if they couldn't eat?

cc

Well, this is presumed to be a long kind of co-evolution process. When you think about the origins of symbiotic interactions, typically you think of two (or more) separate organisms that come into contact with one another. For a symbiosis to develop, the organisms involved must have some reason to stay together; that is, at least one must benefit from the interaction. Over time, the partners adapt to the interaction(s), possibly developing an obligate requirement for each other, as is seen in the case of these mouthless and gutless vent tubeworms and their bacterial symbionts.

ah-d

OK. So then what you're saying is, thousands of generations down, they found it was more efficient to live on what the bacteria were giving them than with their own mouth parts. Don't we have loads of bacteria in our guts? People have known about symbiosis for ages; Darwin probably knew all about it.

cc

They have known about it, but it's often like relegated to a small corner of introductory biology texts, which is astonishing to me because it has had such a powerful influence on the evolution, the physiology, the ecology of all living organisms. I keep saying we are not alone. It's true: everything lives in symbiosis. We share our guts, our mouths and our skin with our microbial symbionts. Indeed microbial symbiosis was critical to the development of plants and animals, with cellular organelles such as the chloroplasts of plants and our own mitochondria having evolved from free-living bacteria!

ah-d
Bacteria are important, aren't they? There are lots of them about.

cc
Yes, there are. The numbers of bacteria found in natural samples are just staggering. Bacteria that are in water and soil are often so tiny you cannot resolve them with a light microscope, but by using special fluorescent stains you can count them. And the astonishing thing is that there are a million bacteria in about a quarter teaspoon of water – seawater, fresh water or even the tap water in our labs, and a billion per gram of soil, a pinch of soil.

Microbiologists knew that they couldn't culture all of them, but now we realise we typically can only culture between 0.01 and 0.1 percent – a tiny fraction. This may be because they're in a quiescent stage, and we're not providing the right media. Or we're simply not mimicking the media or the signals they may be getting from other bacteria. This is one of the Holy Grails of microbiology, to figure out who's there and what they are doing. Molecular approaches now are beginning to be useful, because you can amplify and sequence genes and start comparing them to known organisms.

ah-d

We tend to think that bacteria are bad things: they give us diseases and so on. But clearly you love them.

cc

I do love them. The majority of bacteria are beneficial, indeed they are major contributors to all of the Earth's cycles, e.g., the carbon and the nitrogen cycle. Further, your natural microflora in your gut are protecting you from bad bacteria, germs, because they actually produce anti-microbial compounds that help keep away invaders. It's been estimated that your 'personal' bacteria – your symbionts – outnumber your human cells by ten to a hundred times. Indeed everyone's faecal material has ten or a hundred billion bacteria per gram.

ah-d

It's absolutely mind-blowing, this world you're opening up. So where are you going next year?

cc

We are continuing to explore the occurrence and role of symbiosis in the marine environment. While the vents are a fantastic example of an ecosystem driven by chemosynthesis, given that sulphide and oxygen occur in many other habitats, we proposed that the sort of symbiosis that I discovered in the vent tubeworm should be widespread in nature. Indeed, I found it again in a small clam that lived in mudflats right off the coast. This clam was unusual in that it had a very small gut – indeed, a relative was gutless – and the way that it fed was not understood. It turns out that, just like the tubeworms at vents, these clams had masses of bacteria within their tissues, and depended on these symbionts to make their food. The clam, which lives in Y-shaped burrows, pumps water from the sediment below to obtain sulphide and from the overlying sea to obtain oxygen, providing the bacteria with the chemicals it needs to make energy and food. The cool thing was that this was just the beginning! We and other researchers then re-examined the other vent animals – the mussels, the clams and the tube-worms in the Pacific, as well as the shrimp in the Atlantic and Indian Oceans – all of the major vent macrofauna rely on symbiotic bacteria as their major food source. Thus the symbiotic bacteria benefit by having the animal facilitate access to the chemicals they need to make energy and food, and the host animal obtains its nutrition from the symbionts.

ah-d

You seem to be causing an awful lot of trouble here. You go and look at new animals, and you discover whole new branches of science. Do people love you for this or do they hate you?

cc

Well, I think what it did was to open up a new area. Zoologists actually went back and re-examined shallow-water animals they'd been working on and, lo and behold, found that it was there. This sort of symbiosis has been described now for over 200 species, from six phyla, including these tube-worms, the molluscs, oligochaets and nematodes and even single-celled ciliates. So, researchers all over the world have found this.

ah-d

And I gather you're now not satisfied with life on earth any more.

cc

Well, when you start seeing an ecosystem that is supported by chemosynthesis, it makes you realize that there are other possibilities. I mean before we just always thought of photosynthesis-based life, the way you and I have to eat. Since the discovery of whole ecosystems supported by chemosynthesis in the deep sea, I've been involved with NASA and thinking about life on other planets. When you start looking at Mars right now …

There is evidence of water on Mars and on Europa, one of the moons of Jupiter, but assuming that life requires liquid water, if it is there it is underground – so, how could you have photosynthesis? But there could be chemosynthesis, and that could be the basis for life on these and other planetary bodies. Europa is covered by ice, with geological features that suggest there is a liquid ocean underneath. I went to a NASA workshop and listened to those planetary scientists describing Europa and the convection patterns of the ice, and it was just like being at the hydrothermal vent seminar. It was amazing. There are arguments about whether there's a lot of ice and a little hot water, or a lot of hot water and a little ice. The critical thing is that, in addition to water and a chemical energy source, you need to have something like oxygen or some other electron acceptor for chemosynthesis to occur, as well as a way to regenerate these chemicals because otherwise it'll just go downhill.

ah-d

So, is that your next mission, then?

cc

Well, we are thinking about the origin and evolution of bacteria to guess at what types of metabolism would be possible. I'm not sure that I'll get there to sample!

But back on Earth this year we're going back to a vent site off the coast of Mexico on the East Pacific Rise, where we will be looking for free-living versions of the symbionts in the environment. This may be looking for a needle in a haystack. But, by combining powerful microscope techniques with molecular probes for these individual species, we're hoping to be able to find them.

ah-d

It sounds … well, it does sound like a needle in a haystack. On the other hand, it does sound like a very pleasant needle in a haystack. I presume you spend the morning lying around on the deck …

cc

Well, there was one cruise where we only had two dives, and one of them was aborted. So we were at sea a week, and somebody got a picture of me reading a magazine in a bathing suit on deck. But it's not usually like that. The sub goes down at 8 a.m. typically, and comes up at 5 p.m. So you're down there all day, and you come up right when dinner is served. You don't have time to eat; you are working on the samples all night. And if you are lucky you may get a couple of hours' sleep and then you get up, the sub goes down again, and you repeat the cycle.

ah-d

You're trying to make it sound like a really tough life … But I must admit you also make it sound fantastically exciting. And I do hope your next expedition goes well.

chapter FOUR

Richard Dawkins

Evolution and plain writing in science

Richard Dawkins was born in Nairobi in 1941, his father having moved from Nyasaland, where he was a colonial civil servant, to Kenya to join the Allied Forces. His family returned to England in 1949. He studied at Oxford University, staying there for his doctorate in the field of ethology. He became an Assistant Professor of zoology at the University of California at Berkeley, and later a lecturer at Oxford University, where he was also a Fellow of New College from 1970. His books are *The Selfish Gene*, *The Extended Phenotype*, *The Blind Watchmaker*, *River Out of Eden*, *Climbing Mount Improbable*, *Unweaving the Rainbow*, and *A Devil's Chaplain*. He was appointed to the newly endowed Charles Simonyi Chair of Public Understanding of Science in 1995. He is married to the actress and artist Lalla Ward, and has a daughter by a previous marriage. He won the International Cosmos Prize in 1997, and the Kistler Prize in 2001.

ah-d

Richard, you say that we live for only a tiny piece of the enormous life of the universe, and that we should spend our lives trying to understand why we're here and indeed why the universe is here. But surely science can't answer these 'why' questions, can it?

rd

There are many meanings of the word 'why'. To me, as a scientist, it can mean two things. One is, what is the sequence of events that leads to us being here and being the way we are? And science answers that. The second kind of why is, what is it for? Science can't answer that; indeed, I think it's a meaningless question, it's not a question that should be asked, except for those cases where we're dealing with something that's been designed by people. You can say what a corkscrew is for, what a fountain pen is for, but you can't say what life is for, or what a mountain is for, or what the universe is for. Living things are a special case: you can ask what a bird's wing is for, what a dog's tooth is for. That has a special meaning within the context of Darwinian natural selection; it means, specifically, what has that thing done to help this creature's ancestors survive and reproduce? That's a special kind of why.

But in the colloquial sense of why, I think it's perfectly reasonable to say, why are we here? Meaning, what is the sequence of events, what is the set of antecedent conditions that leads to us being here? I can't imagine a better way of spending my brief time in the sun than trying to understand how to answer such questions.

ah-d

The way we perhaps all know you best is through your book *The Selfish Gene*, which is what, 25 years old now? Are you surprised at how successful that book was and is?

rd

When I was writing it, I did jokingly refer to it as 'my bestseller', not really thinking that it would be. And it was never a mega-bestseller in the first six months, the way one thinks of a sort of blockbuster bestseller, but I've been pleased by the way it's been selling steadily in all the years since. That, in the long run, is a better sort of bestseller than the sort that sells madly in the first six months and then is never heard of again.

ah-d

When you were writing it, were you just telling a simple version of Darwin's ideas of evolution?

rd

I thought I was doing that at the time, and in many ways I still think I was, but I believe there's a difference between popularizing, which means taking something that's already familiar to the scientific community and making it comprehensible – and I've certainly done a fair bit of that – and what I like to think that I've been doing, which is a bit more 'changing the way people think' – and that includes not just lay people but my fellow scientists, my colleagues. I've been told (often enough that I believe it) that even scientists in the field have changed the way they think as a consequence not of anything I've discovered but of my way of putting things, which was sufficiently unfamiliar that it really did turn upside down the way people thought about familiar things.

I've been told (often enough that I believe it) that even scientists in the field have changed the way they think as a consequence not of anything I've discovered but of my way of putting things.

ah-d

Why did you choose the gene as your unit?

rd

I don't think it's right to say that I chose the gene. Choosing the gene was done for me by nature. What I was doing was taking the existing neo-Darwinian theory and saying, this is a gene's-eye view; that's what some people hadn't quite realized. They had been focused on the level of the individual organism. The individual organism was the agent of life: the rabbit, or the elephant, or whatever it was. It's true that the rabbit works to survive and reproduce, but if you ask why the rabbit works to survive and reproduce, what's actually being reproduced are its genes.

What happens is that over many generations the genes that are good at making rabbits are the ones that are still around. Therefore, when you look at rabbits, you see that they are made by genes that are good at making rabbits, genes that have come down through many generations. The gene is the only part of the rabbit, or the elephant, or the human, or any other creature you care to mention, which goes on from generation to generation, in principle indefinitely. In principle, the information in a gene is immortal, and what that means is that good ones are immortal, bad ones are not, and so the world becomes filled with the good ones. Those are the ones that survive; that is the Darwinian process. But individuals, individual bodies, individual rabbits, elephants or humans, they die anyway; they don't survive, not in the long run. Information does survive, and DNA is the information that living things have.

ah-d

You're taking a sort of science fiction view.

rd

It's funny you should say that; in a way, I'm just re-expressing what's in neo-Darwinism, but it is a sort of science fiction view. To describe human beings I used the phrase 'lumbering robots', which caused a certain amount of controversy. It just means that the genes are the information; they're the bit that passes down through the generations. What they use to survive is the body, and the body can be thought of in this sort of science fiction way, as a robot that the genes build for themselves and then ride around inside. And it's because they ride around inside their robot that the survival of the robot is intimately bound up with the survival of the genes. It's a machine that carries its own blueprint around with it, and therefore if it survives and does its job of reproducing, the blueprint will survive.

ah-d

Your thesis essentially says that it's gene survival that matters. But genes have to cooperate at some levels, don't they, even in one individual, let alone in a species?

rd

Cooperation is immensely important. I'm not talking now about cooperation between individuals, which is also important, at least in certain kinds of animal, but cooperation in genes is immensely important, because it's meaningless to talk about single genes in the business of building a body. The business of building a body, which is embryology, is a hugely cooperative enterprise, but it's wrong to think of the cooperative unit of genes as a kind of unit that goes around in time together. They don't – they are constantly split up, separated, and recombined in sexual reproduction. But they do cooperate within each body in making that body. The way to think about it is that those genes survive best which are good at cooperating with the other genes that they're most likely to meet, and that means the other genes that are in the gene pool of the species. The gene pool of a species means all the genes in a species; you can call them a pool, because they're constantly being mixed up, stirred up together in sexual reproduction. The genes in the gene pool build the bodies characteristic of that species, rabbit bodies, elephant bodies, or whatever it is, and the ones that survive in the gene pool are the ones that are good at collaborating with the other ones in the gene pool, to make rabbits or elephants.

ah-d

So you mean as well as being selfish, they also have instructions to cooperate with other chaps? You make it sound simple, but it really is difficult to get hold of this idea, isn't it? People generally find it hard to understand Darwinism.

rd

Yes, especially in the United States of America, and I don't know why that should be. Polls suggest that some astoundingly high percentage, maybe as high as 50 per cent, believe that life, indeed the earth, came into existence within the last ten thousand years, that the Book of Genesis is literally true, and so on. This is just shattering to a scientist anywhere. It goes not just against biology, but also against what we know of geology, of physics, of astrophysics. The whole of science is simply thrown out, in complete ignorance, because it appears superficially to conflict with an ancient book – not actually all that ancient, a Bronze Age text – handed down from a tribe of Middle Eastern herders.

ah-d

So what should we do about this?

rd

Well, education is the obvious answer. I think the scientific community needs to take its educational responsibilities more seriously. It's not enough just to get on with our science. American scientists are second to none in the world, and that applies to evolution as well as many other sciences, but it never occurs to some of them to worry about the fact that they're living in a community in which perhaps as many as 50 per cent don't believe in what they're doing, and believe in something outlandishly ridiculous instead. I think that American scientists and indeed British scientists have got to get out there and start to take their educational responsibilities more seriously, perhaps devote at least some of their time to writing in newspapers.

ah-d

You are in fact a professor of the public understanding of science. How on earth do you profess the public understanding of science?

rd

You could imagine doing research on public understanding of science, actually finding out how many people think that the world is only ten thousand years old, how many people think that humans co-existed with dinosaurs, how many people think that the sun goes round the earth – which, again, is a frightening number. Or you could teach scientists how to communicate.

I actually haven't done either of those. I am doing what the benefactor, Charles Simonyi, himself wanted, which is devoting myself to promoting the public understanding of science – writing books, writing articles in newspapers, giving lectures all around the world and so on.

ah-d

You have said we should spend a chunk of our lives trying to understand why we're here, and you also say we should have a sense of wonder: we should look at science in the same sort of way as we listen to music. Surely you can't do that? You need years of training before you can understand the language of science.

rd

The point about the analogy with music is that we need people to *play* music, and they need years of experience, they need to spend hours a day practising their instrument or they won't play it well enough. In the same way, practising scientists do need years of training. But you can enjoy music, appreciate music even at quite a sophisticated level, without being able to play a note. Similarly, I think you can appreciate and enjoy science at quite a sophisticated level without being able to do science. I want to encourage people to treat science in the same kinds of way they would treat music or art or literature: as something to be enjoyed, not at a superficial level, but at quite a deep level, without necessarily being able to tell one Bunsen burner from another or integrate a function.

ah-d

But isn't language an impossible barrier? I can go and switch on the radio and listen to music and enjoy it, and I don't need to know what it is, who's playing, who the composer was, any of that – I can just hear the noise.

rd

You might be surprised; you need to acculturate yourself to the music you're listening to. We are brought up in Western music, and so we do find it easier to understand because of that.

ah-d

But if I listened to, say, Japanese music, I might not instantly understand, but it would be interesting to listen to, whereas if I went into the coffee room in the zoology department at Oxford – and I'm interested in science, I understand quite a lot – I probably wouldn't understand what two people are talking about.

rd

Two people in the coffee room in a scientific department, talking about research, are indeed using a sort of shorthand language, which they use for the sake of brevity. But it isn't very difficult to switch from that language to one that can be understood much more widely. I feel I have a mission to persuade my scientific colleagues to write their science as if they had a lay person looking over their shoulder, not to write in a language which is completely opaque to other people. I believe they'll do better science if they do that, I think they'll communicate with other scientists better if they do that. I even think they'll understand better the science that they themselves are doing.

You have to get your mind around enormous quantities of time, otherwise you just can't believe that you could go from a bacterium to a human.

The other thing is that different sciences are not easily intelligible to other scientists. I don't understand physics very well, and I think I'm right in saying a lot of physicists don't understand biology very well, though it's probably easier to understand biology. There are aspects of especially modern physics, quantum theory in particular, that are exceedingly hard to understand, totally counter-intuitive. Many physicists have said that they don't understand it either; they do the maths, and they use their mathematics to predict results, and they find that the predictions are fulfilled with astonishing accuracy. The predictions of quantum theory are said to be fulfilled with an accuracy equivalent to predicting the distance between New York and Los Angeles to within the thickness of one human hair. That's what quantum theory can do, and yet many of the people who do it would agree that they don't really understand it at a gut level, because it is so counter-intuitive. If *they* can't understand it, it's not surprising that you and I can't either.

Nevertheless, there are books that really make a strong effort to explain quantum physics, relativity and, in biology, evolution. Evolution has its own difficulties of being understood. It's not as difficult as quantum theory, but there are difficulties. You have to get your mind around enormous quantities of time, otherwise you just can't believe that you could go from a bacterium to a human.

ah-d

William Paley said, if I walk along and I pick up a rock, it's not surprising, but if I pick up a watch, I cannot believe that this was not designed by a watchmaker. What's wrong with Paley's argument of a designer?

rd

I've satirized it as the Argument from Personal Incredulity. But you've only got to think about it a little bit and you realize that there's an infinite regression. It's actually not an explanation at all to say that the watch had a designer, because the designer himself needs an explanation.

That's what's elegant, that's what's beautiful about the Darwinian explanation for living watches: for eyes and elbows and hearts. Because we start from very simple beginnings – simple things are by definition easier to understand than complex things – and we work up by slow gradual degrees, over many, many generations; and because at every step of the way it's comprehensible and the explanation really works. We end up with a complete and satisfying explanation for where the complexity, for where the living watches come from. By living watches I mean eyes, heart, ears and so on.

We understand where they come from, and we don't need to postulate anything mystical or mysterious. The problem with Paley's explanation, with the designer, is that it explains precisely nothing, because the designer himself is presumably even more complicated than the watch, even more complicated than the heart or the eye. So it's a non-explanation.

ah-d

Now, many great thinkers have one thing in their life that they're most proud of. James Watt, who designed lots of steam engines, was most proud of the parallel motion, a lovely bit of applied mathematics. What are you most proud of having found, discovered, invented?

rd

I now think that the book I am most proud of, as a book for lay people, is *Climbing Mount Improbable.* But as a contribution to knowledge I am most proud of *The Extended Phenotype.* It was my second book and it was, as you might guess from its title, not primarily aimed at lay people, though a lot of lay people have read it. It's aimed at my professional colleagues, and I'm most proud of it because the idea goes beyond what others had already done. A phenotype is the manifestation of a genotype. Genes are DNA, a phenotype is something like blue eyes or red hair – or you could think of it even as the whole person. It's that which is manifested. When talking about a particular gene, you would talk about the phenotypic expression of the gene. You would say that this particular gene on the chromosome produces a big nose, the phenotype. So, the genes in me produce my phenotype and the genes in you produce your phenotype …

ah-d

... which goes as far as my fingertips, but no further.

rd

Yes, that's right for conventional phenotypes. But *The Extended Phenotype* says that genes can have a phenotypic effect outside the body in which they sit. I argue this in a kind of step-by-step softening-up way, starting with animal artefacts, with things that animals make, like birds' nests or caddis larvae houses. Caddis larvae are little insects that live in streams and build houses for themselves out of stones or sticks or snails' shells or leaves, depending on the species. So the outer shell of this insect is not part of its body; it's made by the animal. The stone ones are the nicest, because they really do cement the stones with cement that they make. You can watch the insect building; cementing little bricks into a stone wall with great skill.

That is part of the extended phenotype of genes in the caddis larva, and the justification for it is a Darwinian one. It's clearly a Darwinian adaptation. It's well designed in just the same way as the shell of a snail or the beak of a parrot is well designed. It's a product of natural selection.

Natural selection works only by the differential survival of genes, so there have to be genes for caddis larvae houses of various shapes. There have to be good houses and bad houses, and natural selection favours the good ones. It is genes that determine the improvements in the houses – you could say improvements in the building behaviour – but we're accustomed to the idea that when you talk about a gene for something, for some kind of phenotypic effect, it's already at the end of a long chain of causation. The only thing that a gene is really for is a protein. That protein then interacts cooperatively with other proteins made by other genes to produce a complicated sequence of embryological events, to produce the building behaviour – like a recipe when you're cooking something.

Well, if you're going to say the genes are genes for building behaviour, you might as well go one step further and say they are genes for the house that is built. So the house is an extended phenotype. If you accept that, you could say, what about a bird's nest – or maybe a bower would be a better example. A bower bird is a bird that lives in Australia or New Guinea; the male attracts the female not by having a gorgeous tail, as in a peacock, but by building a kind of external tail, which is a bower made of grass, decorated with coloured berries or flowers or Foster's beer cans. This bower is what lures the female. It's not a nest, it's not a house that you live in, it's an external tail attracting females. It's an extended phenotype. They're clearly shaped in this case by sexual selection, and the shaping has to have come about by the selection of genes. That means there have to have been genes for changes in the shape of the bower – the extended phenotype again.

Now imagine a parasite, say a tiny worm, living inside a host, say an ant. The worm manipulates the ant's behaviour for the worm's benefit. Like many parasites, the worm needs to get out of this host, the ant, into the next host of its life cycle, which is a sheep. And in order to do this, the sheep has to accidentally eat the ant.

The worm makes the ant more likely to be eaten by changing its behaviour. It makes the ant climb up to the top of grass stems, whereas the ant would normally go down to the bottom of grass stems. The ant is manipulated like a puppet, like a lumbering robot. The worm sits in the brain of the ant, and makes a lesion, or changes the ant's brain in some way so that the ant no longer goes down to the bottom of grass stems, but goes to the top of grass stems. So the worm gets itself eaten in the lumbering robot that is the ant.

That too is a Darwinian adaptation. I'm entitled to say that because by the normal Darwinian logic we always say that Darwinian selection favours a phenotype, and as a consequence of that, the genes for making that phenotype survive. In this case, the proximal phenotype is in the worm, but the phenotype that really matters is the change in behaviour of the ant.

So now I've softened you up. By going from the caddis, where the extended phenotype is an inanimate house made of stones, we've now got the 'house' being a living ant, but the logic is the same. So the extended phenotype is now allowed to be a living creature, but it's not the living creature whose cells actually contain the genes, it's another living creature.

... to explain something is not to destroy the beauty. In many ways, it enhances the beauty.

The final step, having softened you up so far, is to take a parasite that doesn't live inside its host, and the best example of this is a cuckoo. The cuckoo nestling, the baby cuckoo, manipulates the behaviour of the foster parents so that they feed it. Birds have even been seen flying to their own nest and then diverting to another nest where they see baby cuckoos and they feed them, because the colour of the gape is so irresistible. So now, instead of a worm sitting inside an ant and manipulating its behaviour, we have a baby cuckoo not sitting inside its foster parent, but removed from it by some distance, and manipulating it by light – by the foster parent's sense organs. That too, the changed behaviour of the host, the foster parent, is the extended phenotype of genes in the cuckoo. So it's a long, developing, incremental argument, and you have to read the book.

ah-d
All right, I'll read the book. You obviously love this stuff; you get totally wrapped up in it. You also write about beauty: tell me about the rainbow.

rd
The title of my book *Unweaving the Rainbow* comes from Keats, who complained in a poem that Newton, by explaining the rainbow, removed the magic and removed the joy and reduced it to something dull. I think the opposite. I think, and I think any scientist would, that to explain something is not to destroy the beauty. In many ways, it enhances the beauty. I, like anybody else, like lying on my back in the tropics, looking up at the night sky and seeing the Milky Way; that's a beautiful, transporting experience. And it's not in any way reduced by such knowledge as I have – which is by no means the knowledge of a proper astronomer – about what the Milky Way is. The fact that I know in my limited way that when I look at the Milky Way I'm looking (backwards in time, which is even more wonderful) at our own galaxy, and that there are other galaxies, indeed billions of other galaxies, with the same general properties as ours, only increases the beauty.

ah-d
What's the next book about? Give us a sneak preview.

rd
In addition to a book of collected essays called *A Devil's Chaplain,* I'm writing a history of life under the provisional title *The Ancestor's Tale.* It's a history of life going backwards; it takes human ancestry, starts in the present with us and says, what were your grandparents like? Your great-grandparents? Going on further to your ten-thousand-greats grandparents, your million-greats grandparents, and so on. It's going back to the origin of life, tracing our ancestors; we go through an ape stage, a lemur

stage, a shrew-like stage, a mammal-like reptile of some sort, something a bit like a lungfish – it'll go on back until we hit bacteria. I treat it as a pilgrimage to the past – indeed, that could have been an alternative title, *Pilgrimage to the Past.* As we go on this pilgrimage to find our own ancestors, we are joined successively by other pilgrims. The first ones to join us are chimpanzees; they join us at about six million years back. And then gorillas and then orang-utans – and, not each step of the way, but every few million years, we are joined by more pilgrims, so there's an ever-swelling band of pilgrims as we go backwards, till finally the whole of life has converged on the ancestral bacteria.

There are also tales, as in *The Canterbury Tales.* The tales are things like *The Beaver's Tale, The Fruit Fly's Tale, The Coelacanth's Tale.* Each tale is told by one of the pilgrims, and it is an excuse for describing some sort of general principle of evolution. It's not literally told in the first person, by the beaver or the bonobo, but it's a point of general importance that is best seen at the moment when you deal with the dodo or the bonobo.

ah-d

It sounds wonderful. Do you get right back to primeval soup?

rd

Yes, that is my Canterbury.

ah-d

Now, going back to the public understanding of science, why is it, do you think, that people very often find it easy to believe things that are clearly impossible?

rd

My mind immediately goes to something like what quantum theory requires us to believe, which really does seem to be impossible, yet the predictions are so accurate that we have to believe it. I rather empathize with Lewis Wolpert's book, *The Unnatural Nature of Science*, where he points out that science is actually very counter-intuitive, and that TH Huxley was probably wrong when he said that science is little more than organized common sense. There's a fair bit of organized common sense about it, but there are also major leaps of counter-intuition, major leaps against common sense, which have to be taken.

The thing about ghosts and astrology and other things that people believe is not that they're impossible, but that there's not a shred of evidence for them. People are gullible and credulous, and believe things that don't have any supporting evidence, without bothering to find out the wonderful things for which there is supporting evidence. At any time, of course, somebody could come up with some supporting evidence for ghosts or astrology, but that is no more likely than any of a million other things that people of imagination could dream up. There's no particular reason to pay the slightest attention to ghosts and astrology, any more than to Bertrand Russell's satirical theory that there's a china teapot in orbit around the sun.

ah-d

Are you saying it's actually a problem of statistics or probability, if you like? That people generally don't have an intuitive feel for statistics?

rd

I agree that people don't have an intuitive feel for statistics, and that has consequences. For example, people very often have no idea how to evaluate alleged coincidences, so when they have experiences which they think are telepathic – dreaming about somebody for the first time in ten years and finding out the next morning that they've died – that's a statistical problem, because people don't publish the occasions when they dream about somebody and they don't die.

I've got a theory about this. We calibrated our estimation of coincidence when our ancestors lived in small villages or small bands as hunter-gatherers, where you only ever knew 20 or 30 people, and you gathered round the camp fire and discussed the amazing coincidences that had happened. Back then, if something as astounding as the coincidences that we read about in the *Reader's Digest* or the *National Inquirer* happened, given that our circle of friends was only 30 people, it really would be amazing. But because the circulation of today's newspapers is millions, and it only takes one person to write in with an amazing coincidence, we have a statistically warped, distorted view of how uncanny life is.

The number of opportunities for uncanny coincidences to occur, given that there are millions of people all feeding them in to the newspaper – indeed, the number of opportunities for an uncanny coincidence to happen to any one individual in an average length lifetime – is pretty high. And when you think of all the incidents, when you add up the total number of events in your life, millions and millions of them, it would be surprising if one or two of them didn't have some sort of amazing coincidence-value.

ah-d

You're a neo-Darwinist, is that a reasonable description? Where do you think this branch of science is going in the next generation?

rd

It's moved a long way since Darwin, mainly through the incorporation of first Mendelian genetics and then DNA digital genetics. I'm confident that Darwinism is going to go on, and to be refined, not overthrown. There are some major problems that remain to be solved within Darwinism: one conundrum is, what's sex all about?

ah-d

Yes, I wanted to ask you, why do we bother?

rd

We as individuals bother because that's what we're set up to enjoy. But at an evolutionary level you could say, why don't more species go asexual? Why don't they give up males and pass on all their genes? Theoretically that would seem like a good idea. There's a major sort of industry in neo-Darwinian theory, trying to think of benefits to sex that are sufficiently great to outweigh the enormous benefits of asexual reproduction. Asexual reproduction passes on 100 per cent of your genes, whereas sex passes on only 50 per cent of your genes. The 'why bother with sex?' industry is flourishing; there are various theories going about and people are working on them. So that's one thing that's going to happen in the next century.

Another thing that still remains a riddle is subjective awareness, subjective consciousness. This is a huge problem that is being tackled by philosophers, psychologists and neurobiologists and is, in a way, only peripheral to Darwinism: although whatever the final explanation is that we come up with, it will have to have been a product of Darwinism, if a rather indirect product. It will be Darwinism working via producing brains. And then something about brains will be the more proximal explanatory device that explains consciousness.

ah-d

We've talked about how much trouble people have understanding Darwin's idea of natural selection. Can you suggest some reasons why?

rd

I think it's partly that evolution takes such a long time, but also many people have got the idea that Darwinism is a matter of chance, which it isn't. Chance plays a small part in it in the area of mutation, which is the production of new random variation for selection to work upon; but natural selection is fundamentally an anti-chance process; it's non-random survival, which is why those that have survived are so good at what they do. If it were chance, then any fool can see you couldn't possibly have evolution working. Nevertheless, many people think that Darwinian evolution is a theory of chance, and because they think that, they understandably enough don't believe it.

ah-d

Let us suppose that I don't understand. Can you explain Darwin's idea in 30 seconds?

rd

Every creature alive is descended from an unbroken line of successful ancestors, which means that they have passed down the genes for being successful through the generations. So at every stage, the genes that are unsuccessful are thrown away, by the animals dying or failing to reproduce, and the genes for being successful get passed to the future. That means at any time, when you're looking at the animals that are actually alive, you are looking at the products of a highly non-random sample of genes, ones that are very, very good at building this kind of animal. That's why animals are so good at surviving and reproducing, because surviving and reproducing means becoming an ancestor, and that means passing on the instructions for how to survive and reproduce. That is the survival of the fittest.

chapter FIVE

Loren Graham

Ghosts of Russian engineers and scientists

Loren Graham was born on 29 June 1933, in Hymera, Indiana. He received a B.S. in chemical engineering at Purdue University, followed by a PhD in history at Columbia University. He also did graduate work at Moscow University in the USSR. He is currently Professor of the History of Science at the Massachusetts Institute of Technology. His research focuses on the history of science in Russia and the Soviet Union in the nineteenth and twentieth centuries, as well as contemporary Russian science and technology. Amongst his recent publications, *Science, Philosophy and Human Behavior in the Soviet Union* (1987) examines the state of Soviet research into fields such as psychology, genetics, chemistry, cybernetics and relativity physics; *The Ghost of an Executed Engineer* (1993) tells the story of the engineer Palchinsky's predictions of Soviet corruption and collapse, and simultaneously the story of the Soviet Union's industrial promise and failure, and *What Have We Learned about Science and Technology from the Russian Experience?* (1998) uses the social and economic contrast between the communist scientific community and the West to illuminate the character and social and political status of science and technology throughout the world.

ah-d

Professor Loren Graham has spent his life chasing the ghosts of Russian scientists and engineers. Professor Graham, tell me about Peter Palchinsky.

lg

Peter Palchinsky was a Russian engineer who was born and educated before the Russian Revolution and lived on into the Soviet period. He had a particular view of the way technology and industry should advance, which, in the end, turned out to be contrary to Soviet wishes. He was eventually arrested and shot. But what is interesting to me is the ideas he expressed about the role of technology in society, and how important those ideas are even today, not only for Russia, but also for the world.

He was a mining engineer, a very talented one. He was also what we would probably now call a systems engineer. He lived in Western Europe for a while and he wrote treatises on the ports of Europe such as Amsterdam and London, telling how the various technical, social and economic needs had to be meshed together.

That's what the whole struggle of his life was about. He was an engineer who saw technology in larger terms than the merely technical. He believed that in the final analysis, industry and technology should serve people: and if they don't serve people, then there is no reason to pursue them. When he said that technology should serve people, he meant both society as a whole and the people most intimately involved with it – the people who did the work.

He started writing about this in about 1904 or 1905. The first Russian Revolution was in 1905. He became, I think we would now say, radicalized by the 1905 revolution. He went down to the Don Basin, where much of Russia's coal came from, and he was appalled by the conditions of the workers' lives: the mines, the accident rate, the sanitary conditions of their homes, their living standards. At this time, though, he didn't think of himself as a radical. He just wrote accurate, descriptive reports of the conditions of the mines, for the Tsarist government.

And he found, at first to his surprise and later to his anger, that just telling the Tsar's government the truth about the conditions in these mines alienated him. In fact, he was sent into exile because he told the truth about the mines, even though he was writing the reports on official orders.

Another important part of his approach to technical problems was that good statistics, both about technical things and social things, are absolutely essential if you're going to have a wise policy for industrialization and the development of society. So later, in the Soviet period, when he continued to try to do this, to give them good statistics, he found out once again, just as with the Tsarist government, that sometimes governments don't want to hear the truth.

After the second revolution in 1917, what we now call the Soviet Revolution, he returned from exile and went to work for the Soviet government, and the whole thing repeated itself. They had some suspicions about him because he wasn't a member of the Communist Party, and he didn't have the kind of revolutionary background that they might have preferred, but they recognized his tremendous talents. So they said, help us build socialism, help us build the new state, help us industrialize, get us involved in the first five-year plan, tell us how to build steel mills, tell us how to build hydro-electric dams.

He was good at all this, but he just wouldn't accept their orders and do things in a narrow-sighted way. He would say, well, you want to put a steel plant here? Let's talk about that for a while. In order to make steel, you've got to have coal. Where's the nearest coal? Where's that coal going to come from? What's the freight rate going to be? By the way, you've got to have labour in a steel plant. Where's the nearest populated centre? Maybe you're putting your steel plant in the wrong place. I have another idea. Let's put it over here. And they didn't like that.

This Magnitogorsk steel plant, built in the 1920s was, and remains to this day, the world's largest steel plant. And he said 'Don't build it here', because there was no coal. The nearest coal was a thousand miles away.

ah-d

So why did they put it there?

lg

Because there was a very rich deposit of iron ore there, the so-called 'iron mountain'. Just think for a minute about what the Americans did. The iron ore is in the Mesabi range in Minnesota and in the Michigan range near Marquette. Where are the steel plants? Are they up there where the iron ore is? No, they're in Pittsburgh. Why are they in Pittsburgh? That's where the coal is! Transporting iron ore is much cheaper than transporting coal – and those areas are connected mostly by water, which is the cheapest means of transport.

So Palchinsky said, 'Look, the Americans didn't put their steel mills in Michigan or Minnesota, where the iron ore is. They did some economic calculations. Furthermore, not only is there coal in Pennsylvania, there's also a working population there. The Mesabi range is in the middle of the North Woods. Where are you going to get your workers?' So he asked questions like this, but the Soviet planners didn't like it, because he didn't just follow orders. He kept saying, maybe your orders are wrong. In my opinion, when he said their orders were wrong, he was right.

So, they used him for a while, but he kept getting into trouble. He was arrested several times. He was a very brave man; he could have gotten out of there, but he didn't. He was a patriotic Russian: he even shared the desire of the Soviet government to industrialize and become a great industrial power, so he wasn't subversive in any basic sense. He just wanted to make sure, when they went about such enormously costly activities as building the world's largest steel mill, or building the world's largest hydroelectric dam – he was involved in that also – that they did it the right way.

Eventually they arrested him. They accused him of trying to overthrow Communism and re-establish capitalism in Russia, and they shot him.

ah-d

Why are you so interested in this one man? They shot a lot of people, didn't they?

lg

Well, to me, as a specialist on the history of Russia and the history of Russian science and technology, I've always been intrigued by the fact that the Soviet Union made this enormous effort to become a great industrial power, and in many ways succeeded. At the time of the collapse of the Soviet Union it was the largest producer in the world of a lot of things, including steel, and in that sense it succeeded, it industrialized successfully. And yet, somehow, despite all of that quantitative output – lead, steel, asbestos, you name it – it was a failing economy.

Many people were impressed by Soviet industrialization; it was a model to the underdeveloped world, to countries like India, for many years, before it was obviously collapsing. Why did it collapse when in quantitative terms it was a success? I decided there was something wrong with the way they did it; they had a very narrow view in which what counted more than anything else was just how much of something you produced. They didn't ask questions such as, how much is this serving society, both the workers in the plants and the Soviet Union as a whole? Is this growth in industrial output being accompanied by an improvement in living standards for the nation as a whole? And the answer to that was no. And why? As I looked for answers to that, I found that at the end of the 1920s they arrested a number of Russian engineers for reasons that were never very clear to me. And some of them they shot, like Palchinsky.

I started probing. It took me 30 years to get into his personal archives. They were closed. I kept collecting a little information about him. Then I was there in 1991, a few months before the Soviet Union collapsed; the grip of the secret police was relaxing; the Soviet Union was basically in chaos. So I got into the archives. I started looking at these records and reports that he had written back in the 1920s, where he kept saying to the Soviet government, I'm all for you, build up industry, but don't do it that way! That was really interesting to me. And I think that he became a kind of a key or a symbol, an icon, perhaps, for the failure of Soviet industrialization. I wouldn't say that it all happened because of him, but if you're looking for a single indicator that helps you begin to understand, he serves as well as anybody.

ah-d

I begin to understand; fascinating. Now, you met a Russian engineer. Was it about the same time?

lg

No. I studied at Moscow University. I'm American, I was born in the United States, but the first exchange programmes were set up in the late fifties and early sixties, and I went to Moscow University as a student in 1960–61. I'm a chemical engineer by original training, so I'm interested in engineers, and I noticed that when I met engineers in the Soviet Union and asked them what kind of engineer they were, their answers were different from what I was accustomed to. I was accustomed to people saying, I'm a mechanical engineer, I'm a chemical engineer, I'm an aeronautical engineer, I'm a civil engineer, I'm an electrical engineer. Those are the basic categories in the United States. But the Russians never answered like that.

I remember the moment when it all struck me, as I met this woman engineer. This was at a time when there weren't very many women engineers in the United States, so she was unusual. Anyway, I said, 'Oh, so you're an engineer. I'm an engineer, too. I'm a chemical engineer. What kind of engineer are you?' She said, 'I'm an engineer for ball bearings for paper mills.' I said, 'Oh,' and tried to be polite, and said, 'you mean you're a mechanical engineer,' because that was the closest thing that I could think of in terms of my categories. 'No,' she said, 'I'm not a mechanical engineer. I'm an engineer for ball bearings in paper mills.' I said, 'You don't mean to tell me, do you, that you have a degree in ball bearings for paper mills?' 'Yes.'

... they built factories that polluted, they built factories that didn't serve the workers' needs, they built factories ignorant of economic trends in the rest of the world.

What happened was that after the 1920s, when engineers were educated broadly as they had been in the Tsarist period, the Soviet government completely reorganized the education of engineers in Russia and broke up those broad categories into the narrowest possible specialities. For example, you could be an engineer for either oil-based paints or water-based paints. I'm not kidding; they were broken up that narrowly.

What's more, their educations were not only narrowed technically, but they were devoid of what we would call humanities or social sciences. So they produced the world's largest contingent of engineers. Eighty per cent of the all the people who were educated in the Soviet Union between 1930 and 1970 received engineering degrees. Just think of that. It was primarily a country producing engineers. The result was they built factories that polluted, they built factories that didn't serve the workers' needs, they built factories ignorant of economic trends in the rest of the world. But they did produce steel. They did produce ball bearings.

ah-d
And a canal.

lg
The Belomor Canal. That's another one of the projects of the late twenties that Palchinsky was involved in and heavily advised against. The canal was supposed to allow water traffic to pass between the Baltic and the Barents Sea, through the Arctic area between Finland and Russia. It was built in the late twenties and early thirties by slave labour, prison labour, under terrible conditions. About 10,000 people died every month during the construction.

Furthermore, it didn't really make economic sense. It was frozen half of the year, and the water run-off from melting snow was not enough for the large ocean-going vessels they had planned should use it. It was both an engineering and a social disaster. It's still there. Alexander Solzhenitsyn went there in the late sixties to see just how much it was being used. He spent a whole day on the canal and saw only two barges, going in opposite directions, both carrying timber. The second may have been the same one coming back on the return trip, but still loaded …

There's an important point behind all this, at least to me. Very frequently, when people look back on the Soviet Union, it is said that it was a political disaster and a social disaster and it failed, but people often add that the Soviets were highly successful in building up heavy industry. What I now see is that while they built up industry, they didn't do it in a way that served the people of the Soviet Union or, for that matter, the people of Russia. And now, after the Soviet Union is gone, in most cases they have just had to tear those things down and start all over again.

ah-d
Was that the basic trouble behind the Chernobyl disaster, for example?

lg
Yes. Chernobyl has been discussed many times, and usually the discussions are about the type of reactor and so forth. But I think that the best way of seeing Chernobyl is a little broader than that: as a natural product of the Soviet Union's industrializing, which put primary emphasis upon output to the detriment of everything else. In fact, when the accident occurred, and this is not often noted, the goal of what was being done in that moment was to see how much electricity they could squeeze out of the reactor as they were shutting it down for repair. In other words, the emphasis on growth output was so great – people got promoted by how many things they produced or how many kilowatts of electricity they produced – that even when they had to shut it down for maintenance, they wanted to see how much electricity they could get out of it at the last moment, and that's when it went out of control.

ah-d
So that was partly poor design, but mainly poor policy.

lg
Yes, there's no question it was both of those things. It is an unstable type of reactor. There are still some operating in Russia, but they're being phased out. But just seeing it that way, as an unstable, technically deficient

reactor, is to see only half of the problem. The other half of the problem is that even with those deficiencies, the accident probably wouldn't have happened, at least not when it did, if the Soviet government had not been so insistent on just getting as much electricity out of it as they could.

This was a technocratic top-down system, the engineering-educated politicians dictating everything. The overwhelming majority, in fact about 85 per cent of the Politburo of the Soviet Union at the time of its collapse, was made up of men with engineering degrees. Now, I'm an engineer, so please understand that I'm not slamming the engineering profession. But these were engineers with desperately narrow educations, the kinds of people who don't think in social terms, in environmental terms, but in terms of what we would call maximum output analysis. They've got only one goal, and that is to build something that puts out lots of electricity, or lots of widgets, or whatever it is that the factory puts out. The Chinese are making the same mistake, in my opinion, at Three Gorges – and, by the way, the man who is in charge of that, Li Ping, received his degree in hydraulic engineering from Moscow.

ah-d

I feel sorry for all those engineers; those narrow degrees with no humanities must have been unbelievably boring to do.

lg

Yes, unbelievably boring, and also unbelievably harmful to the rest of the world. I teach at MIT, which is a great engineering institution. Every MIT undergraduate in four years of undergraduate training has to take on average one course each semester in the humanities and social sciences. I wouldn't maintain that that makes a MIT engineering degree a broad education, but compared to what the Russians were doing it's an amazingly broad and sophisticated education, and most MIT engineers have a pretty good knowledge of economics and social change.

ah-d
So the Soviets made a mess of engineering. What about science?

lg
Science is a somewhat different story and a kind of unsettling one, in a way. Soviet science was pretty good in some fields: mathematics and theoretical physics, in particular. I participated in a 1970s study of Soviet science by the National Academy of Sciences in Washington, and among the conclusions were statements to the effect that in some areas of theoretical physics and mathematics the Soviet Union was equal to any country in the world, including the United States.

... if you look at the history of Soviet science, the best work was done in the most awful periods.

Why I say it's unsettling is that we like to think that creativity is associated with freedom. I'd like to think that. It's a pleasant thought that creativity and freedom have a natural association. But if you look at the history of Soviet science, the best work was done in the most awful periods. You've probably heard of the Tupolev airplanes that are still flying. The man who designed them spent years in a Soviet prison. You've probably heard of Soviet rockets. The man who was the leading rocket engineer, Sergei Korolev, spent years in prison. Maybe you've heard of Lev Landau, a great Soviet physicist and winner of a Nobel Prize; the textbook on physics that he wrote with a man named Lifschitz is known to practically every physicist in the world. His best year in terms of publication was 1937. That was the peak of Stalin's purges – people were being arrested right and left

– and even though Landau was very creative in science, he was arrested a year later.

Andrei Sakharov, a great human rights hero and winner of a Nobel Prize, did his most creative scientific work in a camp surrounded by political prisoners. He would look out of the window and see prisoners being marched to work, and yet he was sitting at his table along with Igor Tamm, devising an approach toward fusion energy that is used all over the world, even to this day. In other words, he did a brilliant piece of work under the most incredibly repressive circumstances. So, to make the argument that creativity and science and freedom are inextricably linked, you have to overlook an awful lot of Russian history.

ah-d

Why were they locking up the scientists? I thought they were trying to promote science.

lg

Stalin was extremely jealous of any kind of power that might threaten him and he didn't want a person who was very good at something – even theoretical physics or mathematics or bridge-building or airplane-building – to think that just because he might be the world's best at this, or close to the world's best, that gave him (or in a few cases, her) any authority that could be used to question Stalin's judgement. So, every now and then, he slapped them down. It was inexcusably costly. I may be wrong, but that's how I think Stalin saw it.

The political situation did change a bit during Soviet history. One of the controversies among specialists on Russia concerns the extent to which Lenin and Stalin are similar. The older view, prevalent up until maybe ten years ago, was that Lenin was less vicious. That interpretation is having more trouble these days because we found out that Lenin himself

– although in a somewhat different way from Stalin – signed execution orders for priests and all kinds of other people. So, there's no way that Lenin can be excused from the terror. A lot of people would say there wasn't a whole lot of difference between the two. Solzhenitsyn would say there's no difference.

I do think there was a difference in the following sense. The people whom Lenin executed, he truly believed to be class enemies. He saw orthodox priests as class enemies. He saw Catholics as class enemies. He saw capitalists as class enemies. There is an awful logic here. The revolution has its enemies; we've got to get rid of them. Priests, capitalists, Catholics, those are the enemies of the revolution, according to Lenin. Stalin was different in that he was just about as energetic in executing his friends as he was people who were recognizably enemies.

When Khrushchev came in after Stalin, there was a thaw. I was there during that time. It was definitely a lot better than it had been under Stalin. It was still a police state, but it was a much milder police state. Then Khrushchev fell and Brezhnev came in. Khrushchev still believed in the revolution; Brezhnev didn't believe in much of anything. He was a rather cynical man. He loved his fast racing cars. He just wanted to rule the state, and I'm sure he wanted the Soviet Union to do well in some abstract way, but the revolutionary aura, spirit and élan were gone.

Then, jump forward to 1991, the iron curtain came down and the scientists got their freedom. But to continue the theme that there are many things in Russian history that make us uncomfortable, when the scientists got their freedom, they lost their creativity. Russian science did best when it was in a police state and people were being locked up right and left. Since 1991, it's been a free country to a large degree. No country is completely free, and certainly Russia is not completely free. But within a few limits in Russia today you can say what you want to say, you can do what you want to do, and yet Russian science is doing very badly.

The best explanation is that Russian scientists may have freedom, but they don't have money, which leads to another little uncomfortable polarity. What's more important to science – freedom or money? In two different ways, it looks like money is more important than freedom. During the Soviet period under Stalin, science in Russia had lots of money. He favoured scientists with money. At the same time, he arrested them, locked them up, and treated them very roughly. And yet scientifically Russian scientists did pretty well. Since the disappearance of the Soviet Union, scientists have freedom, but they have no budgets and they're in a catastrophic state.

... if the scientists don't have financial support, freedom by itself won't go very far.

I think this takes away some of our illusions. I wouldn't draw the conclusion that freedom is unnecessary for science. I am convinced that science would do best of all if it had both freedom and money. But we should, I think, be a little more sophisticated than we usually are and realize that scientific research in modern times is a very expensive enterprise. There may be a few areas where you can still do important work on a blackboard or on the back of an envelope, but the number of such areas is much smaller than it used to be and, therefore, if the scientists don't have financial support, freedom by itself won't go very far.

ah-d

Do you think that a particular regime is good or bad for science? Did Marxism get in the way of science?

lg
Marxism definitely got in the way of science in some instances, but it positively helped science in some other instances. First of all, Marxism was a pro-scientific doctrine. Marx described Marxism as a science. And probably no regime in history has ever been more positive towards science in a financial and institutional way than the Soviet Union. Marxism wasn't just a political doctrine to Soviet philosophers; it was also a materialistic doctrine, a philosophy of science, a kind of non-reductionist materialism. Some scientists liked that and actually were motivated by it. The better-known part is that there were some sciences, such as genetics, that suffered disastrously. So the case I would make is that we ought to look upon Marxism and science in a somewhat more sophisticated way than we normally do.

Let's ask the question about religion and science just for a moment. Take the Christian religion, and ask a naive and simple question: is Christianity harmful or helpful to science? Well, the people who want to make the case that it's harmful have plenty of evidence. Look, for example, at the way fundamentalist Christianity has treated Darwin's theory of evolution. Also Galileo wasn't very popular with the Catholic Church. And Giordano Bruno was burned at the stake for his religious heresies. So there are certainly places where the interaction between the Christian religion and science was harmful to science.

On the other hand, any historian of science knows that if you look at someone like Isaac Newton or Blaise Pascal or Galileo himself, you will find that they were quite religious, and in some cases I think that their religion actually helped their science. Newton, for example, had his own very strong religious beliefs. It was not an orthodox religion, but nonetheless it was very powerful, and he believed that his system was showing the grandeur of God's work, the architecture of the heavens. This was

God's work, and it drove him on; no question about that. Pascal had similar thoughts. You can find many scientists, if you're looking for them, who believe they're motivated by religion.

Well, I would say look at Marxism the same way. It's harder for us to accept, because Marxism still makes people uncomfortable. They hear that Professor Loren Graham from MIT is saying, as I am now saying, that Marxism in some cases motivated science, that there are certain scientists who definitely work on the basis of a Marxist approach to reality. They'll say, well, Graham must be a Marxist! But I happen to believe that if you look at the history of science, you can find moments when the Muslim faith interacted positively with science. There was a great age of Arabic science. I can find moments when I think that Buddhism had positive influences on science. Does that mean I'm a Buddhist or a Muslim? No. But I do believe that science can be helped along by Marxist or religious belief systems.

chapter **SIX**

Richard Gregory

Vision, optical illusions and hands-on science

Richard Gregory was born on 24 July 1923 in London. During the Second World War he served in the RAF, where he obtained a scholarship to study philosophy and experimental psychology at Cambridge University. Here he became lecturer on perception, scientific method and cybernetics in the Experimental Psychology department. In 1966 he published *Eye and Brain*, in which he describes basic phenomena of visual perception, explaining how we see brightness, movement, colour and objects, and exploring visual illusions to illuminate how perception normally works – and why it sometimes fails. He founded the international journal *Perception* in 1972, and in 1978 founded Britain's first hands-on science centre, the Exploratory, in Bristol. He is now a Senior Research Fellow at Bristol University.

ah-d

Professor Richard Gregory is a psychologist who for many years has been studying perception. He is a master of illusions, of the way we see, and of what we think we see. Richard, I gather you've just come back from Rome. What did the Pope want to see you about?

rg

Well, I didn't actually see the Pope because he wasn't well, but I went to the Vatican Council, which was set up 400 years ago to advise the Pope. We had a conference on science education, and talked about making the world more rational. It was a lovely mixture of scientists and theologians and we were concerned with science education internationally. A lot of people in the third world get virtually no science.

I was pushing the idea that we get more science centres going internationally, for kids to be brought up with actual experiments. Doing things, trying things out, learning to think on their own, to test hypotheses and then apply science and technology suitable to the country they live in. There were other people there from the third world talking about education as it is at the moment, and so we lived for a week in the Vatican, talking about science and education, and about the status of mankind – not about religion particularly, but about how to live in the world in a way which works, and in a way that is exciting and meaningful.

ah-d

You've always been a great advocate of hands-on science, haven't you? The idea goes back a long way, doesn't it?

rg

It goes back to Francis Bacon, really, in the seventeenth century. In a book called *The New Atlantis* he described an imaginary island whose population could explore the technology and the science that they had, and then do experiments on their own account, and really get away from Greek philosophy. Bacon got fed up with the Greek habit of pontificating and trying to work everything out using logic alone. He thought that people should play about with things, and get ideas like that. The Royal Society was set up from his ideas of how research could be conducted as an enterprise with a lot of people cooperating, which again was a new idea at that time.

But his idea of the science centre didn't really happen until a hundred years ago, or less. The Science Museum in London had an excellent children's gallery in the 1930s, and then, in San Francisco, Frank Oppenheimer set up the Exploratorium. Later, we set up the first one in Britain, called the Exploratory in deference to Frank Oppenheimer.

ah-d
How did you think about the Exploratory? What were your principles?

rg
We followed Frank's principles; in fact I worked with him for a year just after he'd started. One idea was to put perception and physics together so that you got accounts of the world of phenomena, of astronomy and engineering, the actual world of objects. And then you got the human observer. What happens in the brain when we're looking at things? How does the brain construct the virtual reality that is our inner experience of the world?

This relationship between our brain's virtual reality and the actual physical world is a fascinating one. Your mental model of it more or less corresponds to reality, but not by any means entirely. And of course in the case of modern physics, of quantum mechanics, the model in your brain is extremely different from the reality as described by physics.

I think that what the brain is doing is modelling a sort of common-sense physics, a kitchen-sink physics if you like, but not a deep physics, and that this is our inner reality. To be a physicist in the deep sense you have to be a jolly good mathematician and get away from intuition.

So a lot of our experience, our feeling of understanding, is actually wrong, and it gets counter-intuitive when it's right. There's a conflict between what feels right and what is right. Science plays, I think, with the tension between what is intuitively right and what is amazingly surprising.

ah-d

You insisted that the kids in the Exploratory should actually be touching things, that this is how they build up the picture.

rg

Yes, absolutely. It's no good just pushing buttons. You've got to really handle things. Perhaps I could mention how I got on to this idea. I studied a man who was born blind and who then, at 52, had operations on his eyes and got his sight for the first time. Very, very unusual. I found he could immediately see things that he already knew by active touch. Not only coffee cups and small objects, but even telling the time. He'd been taught to tell the time by feeling a hand on a big watch that he had in his pocket. And when we showed him a clock, he could immediately tell the time; we didn't have to teach him.

The same was true even of upper-case letters. The kids had been taught by touch in the blind school, and he could recognize upper-case letters immediately by sight. So we found that the knowledge in his brain from exploratory touch was available to interpret images from the eyes.

But when he hadn't had previous experience of objects to other senses he was functionally blind. These things, bridges and buildings for instance, were like patterns. They were meaningless. They weren't objects to him, they were just shapes. I could tell him it was a building, and then he gradually learnt what it was, but for him it wasn't real.

To start vision going you have to interact with objects. When you're looking at something, you see it's solid. If you look at a glass of water, you know how far you can tip it without it spilling. All that knowledge is in the brain from interacting with other glasses of water, and when you look at it you see these properties that are not visual. So the eye, although it handles only visual information, tells you a lot about the objects in non-visual terms, and that requires learning. It requires a basis of knowledge in the brain from exploring the world.

... it's these gaps ... these failures of prediction which are the key to how the brain and the mind work.

ah-d

You've spent half your life studying illusions – visual mistakes. Wouldn't it be simpler to study the things that are right, as opposed to things that are wrong?

rg

Well, I'm interested in the brain and the mind, and it's when perception deviates from the physical world that you've got a clear phenomenon of mind. If you're looking at a glass of water and your perception exactly corresponds to the object, the temptation is to say that the object simply gets into your brain without any problem, like a photograph. But when there's a difference, when there's a discrepancy, you start to realize that the brain has a problem, that it hasn't quite solved the problem of what that object really is. And it's these gaps, these discrepancies, these failures of prediction also, which are the key to how the brain and the mind work.

In other words, by studying why it goes wrong, when it goes wrong, we then separate out theories of the brain from theories of physics. It's the difference that is so interesting.

ah-d

There's a well-known illusion, a picture of a rabbit or a duck. What's going on here?

rg

What happens there is that you've got information of that object which is characteristic of a duck and of a rabbit, and by tipping it a bit you're biasing it, because rabbits are normally the right way up, so to speak, and so are ducks. So you get clues from the orientation of the ears or the beak.

ah-d

But how can we perceive a solid object as two things at once?

rg

I don't think we do. I think it switches from duck to rabbit. I think of these as hypotheses. In science, like in radio astronomy, you've got signals coming in. You've got a lot of knowledge coming downwards from the brain, the scientist using his knowledge to interpret the signals – either from a radio telescope or from the eye. The analogy here is complete. In order to interpret signals you've got to have background knowledge, and you also need a whole load of rules, like statistical rules and so on, to interpret the signals.

The more likely the object, the less information you need to recognize it. But in an experiment that shows something incredibly unlikely, a sighting of ghosts for example, it becomes more likely that the brain's got it wrong than that you're seeing something veridical. So it's always a balancing of the initial probability of what it is out there and the reliability of the information coming in. You're just balancing or changing slightly the prior probability, the initial probability of it being a duck or a rabbit.

ah-d

You once took me to lunch in a café at the bottom of St Michael's Hill in Bristol, and on the wall there were black and white tiles, all rectangular and all the same size. The lines between them are parallel. Yet it looks obvious that every tile is wedge-shaped. What's going on here?

rg

This is not what I call a cognitive illusion. In other words, it doesn't depend on prior knowledge. This is simply the signals in the eye being upset, from the retina into the brain. Because of the contrast between the black and the white, edges are displaced, producing little tiny wedges, which integrate along, making the big wedges.

There are actually two processes going on. And, by the way, if you take alternate rows and shift them sideways, the wedges then reverse. It's quite a rich phenomenon; you can play with it in many ways. It's near my laboratory which was in the medical school at Bristol, and suddenly one day one of my colleagues noticed it, and then we suddenly realized, golly, that's interesting, and made models of it and did experiments.

ah-d

This café has probably been there a hundred years, and it takes a Professor of Perception to see that there's something very odd going on with the tiles?

rg

That's often the way. It's quite, quite hard to see something very surprising. But it's the anomalies that help us to understand the realities of perception. Not so much the reality of the physical world, but the realities of the processes in our brain for understanding the world. To appreciate how perception works is very relevant to education, and also to art and the artist. How can the artist represent something new or exciting or wonderful? The shock of the surprise is very important; artists such as Escher and Magritte play with surprises.

What the artist in general is doing, really, is representing mind, as well as making us respond to new kinds of surrogate realities, imaginary realities that the artist provides. I think the whole thing's magic.

ah-d

So, I look at a picture, and it's obviously a two-dimensional thing. How does it fool me?

rg

The canvas or the photograph has information of distance. When you're looking at something going away from you, it gets smaller and smaller and smaller. If you're looking at a railway track, the lines converge. In fact, the engine driver has to have a sort of article of faith, because he's driving his engine on something that is shrinking into the distance, and he has to assume that it's really parallel.

So, how the picture works is that the viewer assumes it's really parallel. It's actually converging on the picture or in the eye, when you're looking at the actual railway track. So the brain then says to itself, golly, it's got to be getting further away. It's always a question of whether it's actually converging or whether it's really parallel but vanishing into the distance by perspective.

The artist is playing with that. The picture is paradoxical because although it's really flat, you see depth because it's got cues to depth that are typical of the real world – three dimensions, but presented on the flat plain. So any picture is paradoxical. It's both flat and in-depth at the same time.

ah-d
But on open landscapes there's nothing for a perspective to act on. There are no straight lines.

rg
Well, there are a lot of other cues. It's not only linear perspective, which you get from rectangular buildings and parallel railway tracks. There's the thing of objects getting a bit fuzzy in the distance. More subtly, you get change of distance because everything shrinks as it gets further away, including the texture of leaves on trees and so on.

ah-d
Are we actually fooled? Am I actually fooled into thinking that's a hay-wagon in a stream, or do I say, ooh, that's a nice representation?

rg
I think that one carries in one's mind a sort of double reality. You say, golly, that's a wagon, or that's a bowl of fruit, but you know you can't actually eat it or get in it and ride away with the horse. So you've got this duality. It's kind of 'as if it were a wagon', 'as if it were a bowl of fruit'. You accept it as a kind of virtual or pretend bowl of fruit or wagon. It's like a toy. It's sort of it and yet it isn't it. We live with this sort of double reality whenever we look at pictures.

ah-d
Is that the same as the mental image you're building up in your mind when you're touching things?

rg

When you're touching things I think you build up the non-visual information so that when you look at something, the brain calls upon this knowledge base. And that gives meaning to the visual signal. In the case of a picture, it's a curious double meaning, because you see it as an object and yet you can't actually handle it. It's this tension in art, I think, which is absolutely fascinating.

ah-d

Now, to go back to education. You tell me you've been pontificating in the Vatican; do you think we do things right in this country in terms of science education?

rg

No. I think we want much more hands-on experience, much more letting kids do things for themselves, with guidance of course – you can't reinvent science in each generation. I also think they should learn more about the history of science, about how ideas came about. And, I think, about the human dramas of scientists, some of who succeed and end up with Nobel Prizes and honour and glory. Others, who perhaps are just as clever, just as good, get on slightly the wrong track, and it all goes wrong for them.

There are poignant dramas in science that I think are fascinating. Kids would enter into that just as with a soap opera. In fact I don't know why we don't have some scientists in soap operas on television and radio. They could just gossip in the pub about black holes and all that, about the brain and all the rest of it. Science should be a topic of conversation. It should be part of our culture. And I don't somehow think that school gets kids into the frame of mind, where science is a lot more exciting than magic.

Frankly, I think science is much more interesting than Harry Potter – than magic. A test tube and a microscope and telescope – they're much more powerful than magic wands, much more exciting in what they can do. But kids don't see that.

ah-d
And you think that's because we don't let them get these things into their hands?

rg
Yes, and we don't give them enough of a concept of what's going on. You do need the knowledge in your head to interpret the experiment. Often when you're doing an experiment and nothing happens, that is amazingly exciting, if something should happen and doesn't. So it's not just bangs and whizzes and all that stuff. It's the surprise of the result, whether it's positive or negative, which affects how you think in a very dramatic way. I think if kids had more of a feeling for the drama of this sort of situation in science, they would find the world much, much more interesting.

Frankly, I think science is much more interesting than Harry Potter.

ah-d
You said earlier that you had to form images in your mind, and indeed you're editing the *Oxford Companion to the Mind*. But what is the mind?

rg
I think the mind is what is created by the brain. You've got a physical, physiological, electrochemical system, an amazing bunch of mechanisms in your head – all the grey stuff. And these mechanisms create thoughts, perceptions, colours. When you see the colour red, it's not actually in the

external world; it's created by the brain and generated by the mind, and then projected into the external world. The light isn't actually coloured, it's just a bunch of wavelengths that stimulate the eye and from which the brain creates the sensation of colour. There are wavelengths of light, there is a real world out there, but it's jolly different from how it appears, and the more science develops, the more this difference becomes apparent. The more science gets away from common sense, the more we realize that the reality that science describes is different from appearances.

Another way to think about the mind is that it is like software in a computer. I don't want to say that the brain is a digital computer; it is a kind of computer, though not a digital computer. Take the idea of a physical system that handles symbols, generates concepts and comes to conclusions from data. Now all these are mind, they're the result of the physical system in the brain, in rather the same way that the arithmetic, the word-processing, the decision-making that a computer can do to book your airline tickets, is the result of physical processes handling symbols.

This is the point about the brain: it's a physical system handling symbols.

ah-d

Neurophysiologists look at the brain and wire it up and say that when you're painting, this bit or that bit gets active. Do you go along with all that?

rg

Up to a point. It's very exciting that one can know now which bit of the brain is active when you're (a) looking at something but also (b) imagining it. It turns out that the same bits of the brain are active when you're imagining the glass of water as when you're seeing it. And when you're dreaming, you've got the same bits of brain active even in sleep. That's an exciting discovery.

But I also have reservations about this. For example, if you look at a car engine and say, oh, there's the carburettor, you haven't explained what it does, what its function is, or anything about it. You need to know how the engine works. You need to know what a carburettor does. So, simply to know that when I'm looking at things, this bit of my brain is active, is only the beginning. I need to know what's actually going on in order for the brain to function.

ah-d
If you're saying you get the same bit active when you hold the glass of water or when you imagine it, then you're saying that the mind is not separate from the brain – that it's in the same place.

rg
Yes, very much again like the software in a computer. The processing is physical in the computer, with all the chips and things, but you have to think about mathematics or logic in order to describe what the physical stuff is actually doing. So the study of the mind is like software engineering.

ah-d
You seem to have an enormous range, from kids touching things to the bits of the brain where the software is running. Is this all a continuum or are there different corners of your life that you keep separate?

rg
I think it's all a continuum. But I must say, I've always liked gadgets, even though I'm not a proper engineer by any means.

ah-d
In your flat there's a big room crammed with junk, with Victorian telescopes and zoopraxiscopes. Do you play with these every night before you go to bed?

rg

I do quite a lot, yes. I enjoy them. I like, for example, using an eighteenth-century microscope and getting a feel of what scientists could actually see with it, before they had achromatized lenses; there are coloured fringes around everything.

ah-d

So do you regard science as a great playground?

rg

Yes, I think that scientists play. I think that a laboratory is a kind of nursery, where the public is more or less protected from the scientists, and the scientists get on with what they're doing without being too bothered about administration. This is the ideal, anyway.

ah-d

What are you doing these days?

rg

I'm trying to write a book on perception. I've written several books, about ten I think. But this one is going to be on the cognitive processes linking perception to art, and I'm trying to look at all the phenomena of illusions. I think illusions are meaningful phenomena of the mind, and I'm trying to put them into a structure. I'm trying to organize the concepts of perception using the phenomena of illusions. I'm doing what I call 'the peeriodic table'. It's a terrible joke. It's spelt with two Es, like peering at something you see. So it's a peeriodic table of the elements of illusion.

ah-d

Do you mean there are different sorts of illusion?

rg

Yes. When we looked at that café wall, for instance, the physical signals from the retina were disturbed to create a physiological illusion. Another class is the cognitive illusion, when the signals are misread, because you have to read meaning from signals, and to do that you need a knowledge base. You need knowledge of things like glasses of water in order to interpret the signal. Now, you can misread the signal as you can misread a book. You can get the wrong meaning of an ambiguous word, for example. There's nothing wrong with the word, there's nothing wrong with your eyes. You simply get the wrong meaning from it.

This can happen in two ways. It can be that you're applying the wrong knowledge from handling objects or it can be that rules, such as perspective depth, are not being used appropriately. Indeed, the artist is playing about with this in perspective pictures. He's using the rules by which we see depth and then putting these on a flat plane, which can be disturbing, because the rule is not appropriate to the flat plane of a picture although it is appropriate to the object world. So you get an illusion when the rules are not appropriate, when the artist plays about with the rules.

ah-d

Is there something peculiar in us humans? Because we have art, whereas cats don't paint pictures, do they, or horses?

rg

Indeed, I think it is a deeply interesting point that it's only humans who have representational art and language. Chimps have a very rudimentary language and they can daub away. But it's really only human beings who have both language and art, and who can use symbols in a creative way. This is unique to the human brain. I think animals can make virtual realities, but we can represent them with symbols that we can share in language and in pictures. It's this that's unique, the fact that we can share our mental models, our mental representations. We discuss them, improve them cooperatively, and that gives rise to civilization.

ah-d

David Hockney is a well-known artist who claims that a lot of Renaissance artists were using lenses and things. What do you think about his claims?

rg

Hockney talks about Van Eyck and his famous painting *The Arnolfini Marriage*, which is done in exquisite detail with a chandelier that Hockney said would be absolutely impossible to draw as accurately as was done by Van Eyck without some sort of optical aid. So Hockney's idea is that to transpose the three-dimensional object on to the flat picture plane, Van Eyck actually projected a real chandelier with a concave mirror, and then more or less traced it. It's a bit like cheating.

We know that Vermeer certainly did this. But what Hockney's done is to take it way back to the beginning of figurative painting. *The Arnolfini Marriage* is the first portrait of full-length figures that there is in a domestic scene. So it's really quite early, and he thinks that optical aids were critical for the painter to get it right. This is controversial – it's almost conceivable that Van Eyck was a super, super draughtsman who could do it without an optical aid; so how do we know? I'm inclined to think Hockney is right, but there's the mystery of why the secret was so well kept. Why didn't the sitters for his portrait reveal the fact that mirrors or lenses were used? Why didn't other artists come along and say Van Eyck was cheating? But none of that is in the literature.

ah-d

Fascinating. Now, you've talked a lot about children and their hands-on learning. Is there anything we can learn by watching them?

rg
Oh, a tremendous amount, both for our benefit and for theirs. The mystery and wonder of a child's brain beginning to understand the world through their own experimenting, and through your teaching them, is just wonderful. Any mother knows that, of course, or father if it comes to that.

One thing we can learn is what sort of information kids need and therefore what they need to get it. This affects the toys that we should give children, the education they should get; how much sheer messing around they should have, how much guidance, these sorts of things. We learn what kids need in order for them to see more fully and explore the world more effectively, and then become better, more creative citizens.

Secondly, we can think how we're going to make automatic or artificial minds in robots, in computers, and then how we're going to train them. How do we teach them or program them to see the world? We can learn from how the brain works how to make computers more intelligent. Ultimately we will have computers with eyes that will see things, and will understand the world, I think. They will make decisions, partly because they're programmed. Just as kids are programmed – what schools do is to programme kids – so you program the computer and let it learn for itself.

So, as our technology gets more and more intelligent, we'll be interacting not only with children and other human beings, but also with intelligent machines. And if we can combine machine vision and our vision, we'll have an amazing world where we become brothers to the machines; they'll teach us, and we'll teach them. I like to imagine a post-Bacon world, where civilization will be absolutely enmeshed in intelligent technology.

chapter SEVEN

Eric Lander

Human genome project

Eric Lander was born on 3 February 1957 in Brooklyn, New York. He studied mathematics at Princeton University in 1978; as a Rhodes Scholar he received his DPhil in mathematics from Oxford University in 1981. He then became a professor of managerial economics at the Harvard Business School, where he taught for eight years. During this time, he learned molecular genetics, moonlighting in labs around Harvard and MIT. He became a Whitehead Fellow at the Whitehead Institute in 1986 and Professor of Biology at MIT in 1989. He founded the Whitehead/MIT Center for Genome Research. More recently, he has become the founding director of the newly created Broad Institute, which is a joint venture between MIT and Harvard. The Broad Institute is aimed at realizing the potential of the human genome for medicine. He is now working on the full description of the contents of the human genome, as well as on the determination of genetic variation within the human population and its correlation with susceptibility to disease. He also hopes to clarify the public misconceptions that link genetics to determinism; he sees genomics not as a temporary technology, but as the harbinger of information-based biology.

ah-d

I'm privileged to be able to talk to Dr Eric Lander, one of the key players in the human genome project. Eric, what is a genome?

el

Genome is just the word we use to refer to all of the genes, all of the DNA in any organism. It's the whole set of genetic instructions. There's essentially the same exact DNA in every cell in your body. It's the DNA that went into the embryo that made you, the sperm and the egg brought them in. And they've been copied, three billion letters of heredity from your mum, three billion letters of heredity from your dad. The DNA comes in chromosomes. Each of the 23 pairs of chromosomes in your cells constitutes a single molecule. Extremely long strands of DNA make up these chromosomes.

ah-d

How long? I mean, if you stretch them out.

el

If you were to take all of the DNA from one cell and stretch it out together, it would be about a metre long. If you took all of the DNA in all your cells and stretched it out end to end, you'd get to the moon and back many times.

The entire human genome project is nothing more than sitting down and trying to read this book of instructions, these three billion letters. It sounds kind of easy: you ought to just start at the first page and read the next one and the next one. The problem is, of course, since the letters are really just chemical entities strung together, we have no way to start at page one and just keep reading.

The way we read this text of hereditary information is to take cells, extract the DNA and shred them up at random. We could read only about 600–700 letters at a stretch; so we have to read a three-billion-letter book by reading about 600–700 letters at a time in a roughly random order, and then patch together the story.

ah-d

So it's like taking *War and Peace* and tearing the pages up into little bits.

el

It's many times worse than that. It's like taking a library, shredding the whole thing, and then trying to patch it all back together.

ah-d

Now, when you've unravelled them, do they make sense? Is it a nice tidy set of instructions?

el

No, no. If you went in there as a human editor, the first thing you'd do with your red pencil is throw out all sorts of stuff, because the amount of human DNA that actually codes for the instructions of proteins, like the keratin in your hair or the collagen in your skin, is only about 1 per cent, or 1.5 per cent of the whole text.

ah-d

You mean 99 per cent is rubbish?

el

Well, rubbish is going a little far. But 99 per cent of DNA is not the instructions that make the proteins. Maybe another 1.5 per cent provide instructions that say when to turn the genes on or off. At least 50 per cent of it is what are called jumping genes, transposable elements that copy themselves around and don't have any apparent use.

The way the information is distributed is also really unsettling. It's scattered about. You have a gene on your X chromosome called the 'Duchenne's muscular dystrophy gene', because if it's broken that's the disease that arises. It has only about 16,000 letters of information, but those letters are scattered among two million letters along the chromosome.

So there's a little patch of information here, a little patch of information there, and when your cell wants to access that information it copies a message into the RNA, and that message is two million letters long. Then it clips it down to a mere 16,000, throwing away the rest. So it's like madness. It's no way to run a genome. If you had a smart committee of engineers, they'd undoubtedly clean this whole thing up.

Of course, it wouldn't work. What's striking is that this way of organizing a genome, with the information scattered around in these pieces with lots of stuff in between, is probably very clever. A lot of what makes a genome really work, not just work developmentally in an individual but also work in the form of being evolvable over millions and millions of years, is probably the way that information is packed in an unusual way.

ah-d

Ah, you mean it has to leave room for improvement.

el

Well, it may be room for improvements in some interesting ways. You see, if your chromosomes break and the cell randomly re-combines them, that may be bad for you. But it may be very good for evolution in the long run. And if there are lots of buffer and spacers, then most breaks and sticking things back together may still work, because most of the breaks will occur in the spacers.

So there's an entire theory about how this way of distributing information lets a cell be much more fault-tolerant over the course of evolution.

ah-d

How do you know that those gaps – the interstitial bits – aren't coding for something? I mean, could it be that you just missed the point?

el

This is the fascinating thing about reading a genome. We've gotten the keys to this amazing library with information that evolution has been taking notes on for 3.5 billion years, and we're kindergarteners reading the information. So anything I tell you about what's in a genome – or what its function is – is based on just the smallest experience with what's really in it.

> *You shouldn't think of a genome as a tidy book or a tidy encyclopedia. You should think about it as a churning sea of information, always in a draft form, always being revised over the course of time.*

That said, our kindergarten ideas about what this classic text is about tell us that half of the text consists of repeated elements: elements that copy themselves and have essentially the same sequence, although with slight variations. Most organisms have these transposable elements that are thought of as perhaps selfish genes. Whether they have a good function for you or not,is a tough question to answer. One point of view is that they're selfish and only look out for themselves, and you're merely the vehicle for carrying them around. Another point of view is that they didn't originally have any function, but evolution has come to use them to space things out in helpful ways.

So it may not matter what piece of spacer is there, but some piece of spacer may be needed. And then there are even some thoughts that some of these spacers contribute usefully to your genome by providing places where the cell can start copying DNA more easily. Some of these get taken up and used by the cell. So you might almost think of your genome as a little bit of an ecosystem, where you have your own genes in it, but with lots of other stuff going on.

You shouldn't think of a genome as a tidy book or a tidy encyclopedia. You should think about it as a churning sea of information, always in a draft form, always being revised over the course of time.

ah-d

Sounds like the Internet – most of it wrong, most of it junk.

el

Yes, in a way.

ah-d

Now, you've worked out the entire sequence of bases on this genome. But isn't that just one person who's a sort of schoolteacher in Sri Lanka, or wherever?

el

Well, a guy from Buffalo, New York, actually.

ah-d

But surely this describes only one person. And everyone on earth's going to have a different genome.

el

Yes. We know that any two people – you and I, for example – differ by about one letter in a thousand in our genomes. So we're 99.9 per cent identical. On the other hand, out of a genome of three billion letters, that one in a thousand difference means that there are three million differences between us.

The first job in the human genome project was to seek one single example of the human genome, to get a reference sequence. Once you have a first reference sequence, it turns out to be much easier to get a second and a third sequence. So in fact we've now taken 24 different people, mixed their DNAs together, and sequenced lots of structures of DNA from this big pool. And because we have a reference sequence, we can lower those sequences on to the reference and say 'Ah-ha, here's a change, here's a change, here's a change.' So we now have a compendium of more than two million of the sites of genetic variation in the human.

In fact, in our whole population, there are probably only about eight million common sites of variation. Within a few years I think we'll know not just the reference sequence, but also the vast majority of the sites of common variation. Which brings us to the question of what they do.

We should be able to write down where all the sites of variation are, and then correlate the presence of an A or a T here with, say, your risk of Alzheimer's disease. We actually know some of those already. There's a gene on chromosome number 19 that goes by the mouthful 'apolipoprotein-E.' And it has three possible spellings. People who have one of those particular spellings – it's called the E-4 flavour of the gene – have a much higher risk of Alzheimer's disease.

We know a few examples for things that affect blood clotting or diabetes. But we only know a tiny minority of how these spelling variants correlate with your risk of disease. It shouldn't take long to have the sequence all finished, and to have the list variants all completed. But to correlate the variants with each and every disease is serious work – probably the work of the next two decades or so. The point is there are a lot of diseases. We have to collect patients, and controls, and monitor them, and that will take years.

ah-d

When you've done that, then there are enormous ethical questions, aren't there?

el

You could say we're doing this so we can do diagnostics on people. Or you might say we're doing this because we would actually like to know the mechanism of the disease so that we can fashion therapies directed at the causes, rather than what we have today, which is largely therapies directed at the symptoms.

The problem is that we're going to have the ability to do diagnosis – and pre-symptomatic diagnosis – before we can make any therapies. So I think there's going to be an uncomfortable period where we can make a prediction, like Mr X is at risk of getting Alzheimer's disease, but there's no particular medicine we can prescribe to delay it.

The insurance companies may well want to know about these predictions, which is sinister. But I'm happy right now that the drug companies are using that information to try to make medications that will slow down the process of Alzheimer's, so that perhaps five or ten years from now one can begin to take preventatives that would push Alzheimer's out, say, from your seventies into your hundred-and-twenties …

ah-d

I don't want to be that old.

el

But my point is that this carries with it a lot of responsibilities. Whose information is it? Does it belong to me personally, my insurance company, my employer, the government? We have to come to some understanding about the privacy of that information, because in principle this is great information; it ought to be able to let each person make a choice of whether he or she wants to know it, and how he or she wants to use it. It ought to be able to let the medical community work out optimal therapies.

But only if we protect our privacy. If people are concerned that they're going to lose their insurance by finding this out, they're not going to find out. And if they're concerned that that's going to happen across the board, they're not going to support medical research. In the long run we'd be doing a tremendous disservice to our children if we didn't pursue these things. But we'd better get it right.

... we'd be doing a tremendous disservice to our children if we didn't pursue these things. But we'd better get it right.

ah-d

Going back to the great project, there were two rival genome teams, right – public and private? And the public guys won. That's you.

el

Well, the general public won.

ah-d

If a private company had solved all this, all those questions would become much sharper, wouldn't they?

el

Certainly the public human genome project, of which I'm a proud member, felt strongly that this information belonged freely in the public domain so that anybody could use it. Any one of thousands of biotechnology or pharmaceutical companies, any one of millions of people can access it – with no restrictions. We made our data available to the world.

ah-d
How?

el
We put it on the Internet every day. Every 24 hours the team posted all the new information on the Internet. So it just kept going up and up and up and up. And anyone can use it.

ah-d
And do they?

el
Oh, yes, yes. Tens of thousands of hits on these Websites occur every day. The entire collection of the genetic information has been downloaded to large numbers of research labs – thousands of sites around the world – and people consult it every day. Any problem you're working on, if you're studying a cancer or a brain-degeneration disease, you find yourself constantly going back to the reference table, much like anybody doing chemistry would go back to the periodic table.

Of course you've got to know what the elements are. You're going to look up their properties. Now, the periodic table, the chemists probably memorize all that. That's not possible with 30,000 genes. You really do have to go and check them.

ah-d
That's amazing. I mean, to put out this information every day …

el
One of the greatest things about the project is that when we wrote a paper

in the journal *Nature* in February 2001, we were able at the end of that paper not to do the usual scientific thing of speculating how this work might be used some time in the future, but to include an entire section describing the discoveries that had already been made with the information that had been posted during the course of the project. It was wonderfully satisfying.

ah-d

That's terrific. Now, let me go right back. You're not a biologist at all; you're an interloper in this area.

el

A complete interloper. By training I'm a pure mathematician, completely unqualified to be doing biological research. But I think it says something about what biology is today.

ah-d

And you stepped off into Harvard Business School.

el

Yes. I have a very chequered career path. I went and taught managerial economics at the Harvard Business School for a number of years. While I was on the faculty there, I got interested in biology through a series of wonderful accidents and a suggestion from my brother, who is a biologist. I began reading about neurobiology and thinking about the information coded in the brain, which seemed like a problem in mathematics. And one thing led to another: to understand that, you had to understand cell biology, and to understand that, you had to understand molecular biology, and to understand that, you had to understand genetics.

So I found myself over the course of a wonderful summer doing this regress back to genetics and found it just so captivating, and I found that biology was beginning to make a transition into an information science. When I was in high school it was never like that. It was all flies and cats. Now it's clear that what unites all of biology is this fundamental information system that life invented 3.5 billion years ago with DNA and the way it codes for genes. Because you can look across from a human being to a yeast and find that while they look so totally different, at the DNA level there's a set of genes involved in, say, yeast cell division. And they're the same genes for human cell division. In fact, they're the ones that go wrong in cancer.

If you want to study cancer, you can study lots of it in a yeast. If you have a yeast that has a defect in its ability to divide its cells, you can take the human gene and cure the yeast. Presumably you could also do the reverse, although it would be unethical to do that. The unity of life is so apparent when you look at the genetic level, the information level, that evolution invented things only a couple of times and then re-uses them. Evolution's tremendously conservative.

So, whereas 20 years ago mathematics played no significant role in biology – there weren't large data sets to analyse – we're now in a world where studying biology requires looking at vast data sets – three billion letters of human DNA, three billion letters of a mouse DNA, lining them up and comparing them to see what's the same and what's changed.

Nowadays if you want to study a cancer, you can take a tumour, grind it up, purify the RNA messages from that tumour and wash them over a gene chip that has a detector for every gene. Then you can read out how much each of 30,000 genes is turned up or down. So every tumour becomes a list of the 30,000 genes up or down, and you can see which ones look wrong.

It's a whole new world. And 20 years from now, biology will be populated by a large number of people who think of themselves as biologists, but half of what they do will be computational and half of what they do will be 'wet science' at the bench. They're not going to even think about that distinction any more. It's all going to be one.

They're going to look back on the wonderfully benighted twentieth century and wonder how in the world anybody ever got anything done.

ah-d

There seems to be a big change in progress. Just in the course of these talks I've talked to Bob May, who started life as a theoretical physicist and then went via ecology into population biology. I've talked to Lewis Wolpert, who's an engineer turned embryologist, and so on and so forth. People coming from physical sciences and invading biology.

We're never going to be able to write a set of equations that describe life, because life is far more historical accident.

el

I think that says a lot. Biology is becoming a quantitative and information science. Not in the way physics is; I mean, in physics the goal is to write a set of equations that describe the world perfectly. That isn't the goal in biology. We're never going to be able to write a set of equations that describe life, because life is far more historical accident. The details of the way that a cell works depend a lot more on the historical contingencies of how things happen.

So it's not going to be derivable from first principles. But we're grasping the idea that it's a vast information system and that we can learn a lot about it by stepping back and taking the global picture, the big picture. Instead of studying one component of the cancer cell we can study all 30,000 genes and see how they're working together. Instead of having your favourite hypothesis about what's the basis of some brain degeneration, take a completely unbiased look at all possible genes and let the cells speak to you, let the organism speak to you. In some sense this is how the organism can communicate to us about what's important. That's what's great about biology.

ah-d

The way you're talking about it, with this wonderful evangelical overview, sounds very much like Charles Darwin. He went on a trip around the world in the *Beagle;* he went back to his house in Kent and he sat and thought about it for 25 years; and then he wrote this book outlining his idea of evolution by means of natural selection.

It seems to me that you biologists or mathematicians, or whoever you are, are beginning to take the same sort of global look at biology and analyze it in a more mathematical and more precise way. But it's the same sort of principle.

el

I hope we come up with insights as deep as Darwin did. And of course that's what we're trying to do. We want to look at the cell now as a complex circuit, a complex set of machines. We want to recognize the modules of a circuit, which evolution has used and re-used, whether they turn up in an animal cell, a plant cell or a yeast cell.

The next question will be whether clear principles emerge that provide us not just the details about this lung tumour, but also general principles about what makes a tumour progress. Or even more general principles about how evolution is able to continually tweak the body plan of a mammal to get all the different kinds of dogs there are, for example. I'm hoping those things will begin to become clear in the next couple of decades.

But the first step is to be able to take in this big picture. It's like a blind man and an elephant. He can make some sense and try to solve some problems by feeling the tail or feeling the tusk, but his information is desperately limited. Feeling the tusk he might say, 'Oh, the monster's very hard and pointy.' But if he feels the trunk he will think something quite different. Now we're finally able to step back and see the whole elephant. That's going to totally re-orient biological research. But exactly how, and what we're going to learn from that re-orientation, it's too early to tell.

ah-d

You mentioned a gene chip; I don't understand.

el

Well, it's actually easier than you might think. Crick and Watson worked out that DNA is this wonderful double helix where each strand is complementary to the other. That principle allows you to build gene detectors. If I take one strand of a gene and attach it to a piece of glass, I can then sample at every spot along that piece of glass. Let's say I have a row of little squares, and in square number one I put strands of DNA from the first gene, in square number two I put strands of DNA from the second gene, and so on. Now I have a little strip of glass that's a detector for all genes.

I take the messages from a cell, fluorescently label them so I can follow them, wash them over the chip, and by the wonders of Crick and Watson base-pairing, the message from gene number 17 will find its little detector and stick to it. Then I can just run a laser across it and see by the fluorescence intensity how much of gene 17 had been turned on, just because it's going to stick to its own little spot.

So I can make a chip with 30,000 spots corresponding to each gene, take the cell's messages, wash them across, and the 30,000 spots read out the amount of each gene's transcription, its expression. So every tumour, for example, if you were studying cancer, can be described in terms of which genes are up and down. Researchers have found that what physicians call one kind of lymphoma is really three different kinds of lymphoma, because they all have different patterns of gene expression. The tumours are smarter than we are. But we can ask them to account for themselves. They know what's going on; we often don't. We just have to ask.

ah-d

Arthur C Clarke's First Law says that any sufficiently advanced technology is indistinguishable from magic. That's exactly what you're describing.

el

Of course. Even 20 years ago this would seem like magic.

ah-d

When you started in 1980, how much could a graduate student do in a day?

el

Oh, golly, back in 1980, doing several hundred letters of DNA in the course of a day was a pretty good day's work. Sequencing an entire gene back then was worth a PhD thesis.

ah-d

And now?

el

Our centre, at the Whitehead Institute, produces about 30 million letters of DNA each day. If you had two bacteria you'd like to sequence, we can do that in a day.

ah-d

That's not just because you have graduate students who are better at the job, is it?

el

No, no. In fact, none of this is done by graduate students any more; all that work is done by robotics.

ah-d

So how has this happened? I mean, I don't believe somebody came along and invented a robot. This must be a sort of symbiotic process of man and machine?

el

This has been a steady auto-catalytic process. DNA sequencing technology went from the brilliant ideas of how you could work out the letters of the DNA chain to many chemical improvements, computing improvements, engineering improvements that would allow you to attach fluorescent tags, or run samples by laser detectors, or use computers to assemble the little bits of sequence better. Now we have the robots that can not just mimic what a graduate student does, but use completely new chemistries to precipitate DNA on beads and magnets.

So now when you walk through a human genome centre, it looks very much like a factory. We have conveyor belts in our lab moving around plastic plates with 110,000 samples, and only a handful of people minding the machines.

Almost all the people in the centre are involved in studying the results of the sequences rather than doing the mindless tasks of sequencing. But all this is a result of 20 years' work by hundreds and hundreds of people around the world making sets of improvements.

ah-d

But very focused. Almost never in the history of science can there have been such focus around the world on a single problem.

el

That's probably true – there's been a tremendous amount of focus. But this is typical for biology. The reason biology has undergone such an explosion in the last 40 years is because there's been a marvellous feed-back loop. Every new discovery about life seems to translate into a new tool to make more discoveries. So when we understand what antibodies are about, and how the body's immune system works, we can then make monoclonal antibodies and use them to probe new things.

When we understand how the cell copies its DNA, we can suddenly copy DNA. So each new trick that evolution has worked out over the course of the past three billion years, as we discover it, turns into a tool in our bag of tricks. That's what's been so powerful. That's why we keep speeding up and speeding up. All the really clever things were invented by evolution. We're just sitting here at evolution's feet learning this stuff.

ah-d

Now that the human genome is finished, presumably you can retire …

el

Oh, no. Not even close. We're sequencing the mouse genome. Others are sequencing rat genomes, fish genomes. Because, in fact, to really under-stand what's in the human genome we need to see what parts evolution cares about.

There are two ways to do that. One is to do a huge number of ex-periments attempting to change each base and seeing what the effect is. The alternative is to let evolution do it for you. Evolution gets up every morning and changes a few bases and sees how it works. So by comparing the sequence of human and mouse, we can find bits that we didn't recognize were important but evolution has lovingly preserved.

ah-d

How different is the mouse genome from ours?

el

Well, in terms of the genes, the mouse has virtually the same set of genes. The exact sequence of them is about 70 per cent identical, about 30 per cent variant. The basic gene set is the same. But in terms of the stuff that doesn't matter, it's drifted a fair amount. And the stuff that does matter – the coding sequences of genes, the regulatory sequences – they've been preserved pretty well. So just by stacking the two up on top of each other, we can see which bits matter a lot.

The embarrassing thing is there are maybe a quarter of a million regions that don't code for proteins, but that evolution has preserved. And for almost none of them do we know why. So, hey, you know, we thought we were so smart, now we have to figure out this quarter of a million unknowns in the genome.

ah-d

Ouch! You manage to make it sound amazingly exciting – I almost want to come along and help, though that might set back progress by years …

el

This is the greatest time to be in the game. I can't imagine a better time for a young person to want to go into science.

chapter EIGHT

Lord May

Chaos, AIDS and science in politics

Lord (Robert) May of Oxford was born on 8 January 1936 in Sydney, Australia. He graduated from Sydney University, continuing with a doctorate in theoretical physics and researching that field for the next 13 years. He then switched disciplines to population biology at Princeton University in the United States, where he remained for a further 15 years. In 1988, he moved to Oxford University in the UK and concentrated on mathematical biology. He spent five years as Chief Scientific Adviser to the British Government and Head of the Office of Science and Technology, and is now President of the Royal Society. In 2001 he was created a life peer, Baron May of Oxford.

ah-d

Bob, 25 years ago, you were one of the founders of chaos theory, I believe. Was it really started by a meteorologist with a duff computer?

rm

Yes and no! The origins lie deeper, but essentially chaos marks the end of the Newtonian dream – the vision of the Enlightenment – in a way. What I learnt in school is that if you've got a really simple mathematical equation, a really simple set of rules with nothing random in them, then you can make predictions. And the things we can't predict – roulette wheels, for example – have rules underlying them and equations, but there's too much going on, and it's too complicated to make predictions.

Now chaos can give you equations, with all the mathematical rules, that are almost as simple as anything you can imagine, and yet they can produce dynamics that are as complicated as anything you can imagine. I can give you the equation, I can tell you the starting value to ten decimal places, and yet you won't be able to make accurate long-term predictions. And that flies in the teeth of everything we learnt in school.

ah-d

So, you start with a simple system and wind up with a complicated one?

rm

You start with a simple system but it describes something very complicated. Two strands brought chaos to the centre stage in the seventies. One was in meteorology – a chap called Ed Lorenz – and it wasn't a duff computer. He took a simple metaphor for a meteorological problem, but simple in a particular way. If you're working in the kind of mathematics we nearly always learn in school, with things that change continuously with time, you use differential equations, and you've got to have at least three dimensions and three equations before you can get chaos.

The second strand was systems that are discontinuous in time, like the population of knapweed, or blue tits in Wytham Wood outside Oxford from year to year, then it's different. The time series is 'next year in terms of this year' rather than continuous change. With this system you can have just one equation, and you cannot predict the results.

Think of a number x between zero and one, and take it away from one to give $1-x$. Multiply these two numbers, x and $1-x$, together and then multiply by a constant, and that's the next value of x. An equation as simple as that can give you something that doesn't just look like random dynamics, but is so sensitive to how you started it, that you can't predict the outcome. Those two strands started chaos theory. Lorenz's results appeared in 1963, but about ten years went by before people began to appreciate the idea, because although it was simple for physicists, it wasn't for anyone else.

Chaos theory now permeates whole swathes of physics and chemistry as well as biology.

But that simple equation, and its friends and relatives, equally simple – they got attention focused on chaos in the early seventies, the two things came together in the mid-seventies, and chaos theory now permeates whole swathes of physics and chemistry, as well as biology.

ah-d

You start with a simple system and it becomes unpredictable. Can you run it backwards? If you see a thing that looks like chaos, can you work out what's going on?

rm

Good question. Let me go back to that equation again for the moment. I know everyone says you should never have a equation in a book, but take a number between zero and one – call it x. Then take x from one to give $1-x$, multiply it by a fixed constant and that's the next x. If the constant is between one and three, that iterative process of next x and next x will settle to a constant like intuition suggests. If the constant is between about three and three and a half, it'll describe something that goes regularly up and down. But if it's between about three and a half and four, it'll give these results that look random.

Then, in principle, you might be able to run it backwards – in the jargon of mathematics, plot the next x against the first x and reconstruct the equation. But remember, even if you know the equation or if you've deduced it from the time series, if you make tiny errors in how you start, things are so sensitive in this chaotic region to how it started that the tenth or twentieth decimal place will have you all messed up by the time you've made ten or twenty iterations. So I can tell you the rule and I can measure the starting condition to what seems to be great precision, and yet, in the poetic image that Ed Lorenz gave us, the fluttering of a butterfly's wings will carry the hurricane to a different place.

And it's not some mathematical abstraction. When I was young, we thought that you could work out the equations that govern fluid flow – what goes on in the atmosphere. The equations are complicated, but in physical terms they're not that bad. As we got more and more fancy computers, we thought we would be able to predict further and further ahead for the local weather. Now we know that's not true. No matter how accurately you measure the temperature, the wind speed, the humidity, no matter how many weather stations you've got, the tiniest errors when you're in that chaotic regime mean a local weather prediction that will be way off the mark. As Tom Stoppard has said, 'Whether it will rain on auntie's garden party two weeks from now will always be inherently unpredictable beyond about ten or twenty days'.

ah-d

Do you listen to the weather forecast?

rm

That's different. The weather forecast tells you what it's likely to be for the next few days, and statistically it's fairly good. The same kinds of argument apply to things like marginal rates of treasury bonds, or currency exchange rates, or other things in the market. You can ask, are these apparently random things? Are their fluctuations chaotic? Or are there, at least to some extent, simple rules – maybe trading rules, maybe computer rules – that are endowing them with a degree of volatile dynamics? Because if there are, they won't be any more help for long-term prediction than reading the entrails of a sheep. But such new 'chaos theory based' approaches can give you a new route to short-term prediction, and there's quite an industry in that.

ah-d

You started life as a physicist, but you switched to ecology and then immunology. You've been all over the shop. Isn't this rather unusual?

rm

I have a short attention span!

ah-d

Hmmm! You're clearly a mathematician at heart.

rm

I'm not a mathematician's mathematician. I'm a theoretical physicist – an applied mathematician. I went to university to become an engineer, and while I was at university I discovered there was this wonderful world where you can spend your life as a researcher. I always liked playing games; I've played chess, I've played bridge, and I played snooker a lot when I was at university. But I discovered there was this hedonistic life where someone was willing to pay me to spend my life playing games – in a sense playing games with nature, where the name of the game was to try to work out what the rules are.

And that's what I've done. I was taught to believe this wonderfully arrogant attitude of the theoretical physics of the golden age, that I didn't have to be a specialist in plasma physics or low-energy nuclear physics – I could think about any problem. This was in the early sixties, as the computer age was beginning and things were mainly being addressed by large teams of people focused in particular areas. The romantic age had given way to a much larger enterprise; so it was in some ways an inappropriate education, but it was one that suited my style, and I was lucky enough to blunder into, by pure accident, problems in ecology and population biology that suited my style of theoretical physics.

ah-d

Why was a theoretical physicist even reading about ecosystems?

rm

Well the late sixties was a time of great ferment on campuses. It was partly a product of the Vietnam War, it was a kind of questioning – it was a time of feeling that science should be more socially responsible – I was drawn into a movement, one of the founding members of Social Responsibility in Science in Australia, and I was trying to find out what to be socially responsible about.

I read a book by a bloke called Ken Watt called *Ecology and Resource Management*, and he set out an interesting problem. He said, 'Here is one of the things that you read in any ecology text of the late sixties: complicated systems are more stable.' Complicated systems in the sense of those with lots of species and a lot of connections among them. But then Ken Watt himself said there were arguments for this, but it didn't seem to fit with common-sense observation. And when I looked there were unconvincing arguments: one of them was that mathematical models of simple one-predator, one-prey systems are unstable; that's only half an argument.

The role of the lucky accident can never be underestimated.

I sat down that same evening and looked at the corresponding models with many-predator, many-prey systems and saw they were typically less stable, and I talked to one of my biologist friends, Charles Birch, who was the real leader of the initiative for Social Responsibility in Science in Australia – a wonderful person. And Charles, who didn't like theoretical things in ecology, but had the generosity not to impose his tastes on other people, said, 'I think that's probably interesting – write to Ken Watt'.

And after that, one thing led to another. I went on sabbatical next year in 1971. In Britain, I met John Maynard Smith, Dick Southward and other ecologists. I met Robert McArthur at Princeton, who had just learnt he was dying of cancer. Robert McArthur more than anyone else was a person who had recently formulated many of the important questions of ecology: how similar can species be and persist together, for example. And he had formulated them in the sort of theoretical physics idiom, but did not have the particular technical skills to carry some of this work forward. I did a couple of things with Robert and, as I say, one thing let to another.

In particular, I got interested in some of the little equations ecologists had written down to describe how populations tended to increase when they were at low density, and to decrease from generation to generation when they are at high density, as they ate themselves out of house and home, or diseases spread. And it was that that led me into some interesting mathematics. But the role of the lucky accident can never be underestimated.

ah-d

But it was Pasteur, wasn't it, who said that fortune favours the prepared mind? If you'd been interested only in Bondi Beach, you wouldn't have looked at ecology. So, you did all your first training in Australia. What happened next?

rm

I did all my early training through my PhD in Australia. I spent a couple of years at Harvard as a post-doc, the most important event in my life; I met my wife there! And then we were in Australia for about ten, twelve years, during which time I spent about three years away in Britain and at Caltech, and again at Harvard on sabbatical. And then in 1973, shortly after I got interested in biology, it was clear that for the things I wanted to do there would be more interesting communities elsewhere. So for one reason or another I moved to Princeton for about sixteen years, and then I moved to Britain about twelve years ago.

ah-d

And you have had some curious posts.

rm

When I was in Sydney, I was in a thing called the Daily Telegraph Theoretical Physics Department, which could be very confusing. Sometimes people dropped the last bit and I would appear at conferences as Robert May, The Daily Telegraph.

ah-d

You then started to apply these ideas in immunology – you worked on HIV for a bit. Again I don't understand how you could possibly switch from population dynamics to immunology.

rm

Well it's a logical thread if you follow it. I got interested in a variety of ecological problems, and one of the issues that up to the late seventies had not really been looked at much by ecologists was the role of infectious diseases as an influence on the numerical abundance or the geographical distribution of non-human animal populations. There are a few case studies, such as the way Rinderpest has affected the geographical distribution of wild animals in east Africa, for example. But, by and large, my ecological colleagues who work in the field like to look at romantic animals in romantic parts of the world, and they're much less interested in looking down microscopes at grotty little disease organisms.

On the other hand, when you begin to think about it, infectious disease has been a huge shaper, not just of the non-human animal world, but of human history too. It was really with microbial infections that the old world conquered the Americas, Australia and Oceania. Disease is arguably no less important in setting the bio-geographical patterns of plant and animal distribution. So a colleague, Roy Anderson, and I got together with a rather self-conscious programme of saying 'Let's try to understand how infectious diseases can affect animal populations'.

We worked on that in the late seventies, early eighties. Looking for really good data to test the ideas, we found that some of the most interesting information was public health data for infections of humans, particularly cyclic patterns of measles in big cities before the advent of vaccination. Measles had always been a notifiable disease back to 1900; so there is data in England and Wales for cases of measles week by week from the early 1900s up to about 1960, when the advent of vaccination ruined that beautiful pattern. That, in a sense, pre-adapted us, so that when HIV and AIDS came along, we were in a position to work on the data, and we in fact published the first projections of what the probable demographic impact of AIDS in central African countries could be.

We were lucky because we had the sort of ecologists' demographic background, but we also had the more specialized work we'd been doing in looking at the dynamics of infectious diseases in populations in a data-related way. We produced projections that were at the time thought to be very pessimistic; in fact, the World Health Organization and the Population Council in New York City thought our projections were far too pessimistic.

The difference between our calculations and theirs arose because we had been looking at the data about transmission rates, and in particular the connection between the number of sexual acts and the chance of getting infected, in ways that focused on the data, rather than lifting off the shelf a measles-like model and whacking it into a standard demographic model. And much to our regret, in many ways, our models were right, and the World Health Organization and the Population Council were wrong. And it has indeed been as bad as we feared. Roy Anderson went on to pursue this epidemiological approach with more and more resources, as there was more and more demand, and Roy has done many interesting things since.

With my short attention span, I figured, 'been there, done that', and I turned my focus to a still unsolved problem for HIV: why is there so long and so variable an interval between getting infected with HIV and the onset of AIDS? The huge majority of scientific research on AIDS is focused on understanding how the individual HIV virus interacts with cells of the immune system. Many of the people who do that don't lift their sights from that particular specialized problem to the larger question: what is the parthenogenesis of HIV and the human host? Many of those who do lift their sights think that an ever-more-minute description of the interaction between individual viruses and individual immune system cells will lead to an understanding. But my own view is that a part of the understanding – a part only because you need the individual immunological understanding of detail to begin – is also going to involve a clear insight into how whole populations of HIV virus interact with whole populations of different kinds of immune system cells.

And that carries us right back again to a problem in population biology, with populations of cells in the body, rather than populations of lions on the Serengeti.

ah-d

But why is this interval important?

rm

It may be – we don't know. Many of the vaccines we've successfully developed, to the great benefit of public health, have been discovered by a sort of mixture of accident and the search for things in the natural world. Think of variolation with cowpox against smallpox, or the recognition that attenuating viruses stimulate the immune system. So it's worked, and it's worked with the hands of very able and gifted people, but it's still in many ways as much an art as a science.

It may be that we'll be able to develop a vaccine against AIDS, against HIV, without understanding the detailed mechanism of how infection with HIV produces AIDS after different time intervals in different people, but I personally believe ten, twenty, thirty years from now we will design vaccines from the molecules up. Instead of this blend of science and art and happy accident, we will actually ask what is actually going on by looking at individual viruses and individual immune-system cells, and how whole populations of these interact.

And we'll ask – at the molecular level – are there bits of the HIV virus that we should focus attention on? And I think we may then have a clear enough understanding of the process that it will help us design an effective vaccine. I may be right, I may be wrong. But that to me is a reason for wanting to understand the process of exactly how infection with HIV eventually leads to the onset of AIDS. Quite apart from the general reason that drives most scientists much of the time – whether or not there's an application – you just bloody well want to know!

ah-d

Insatiable curiosity! You seem to hop happily from low-temperature physics to the infection interval in AIDS. This is astonishing. Is this because you're really a mathematician?

rm

I'd say it's because I am, in a sense, a theoretical physicist in the classical, romantic age paradigm of theoretical physics. My interest really is in understanding phenomena in the natural world. And as the science fiction writer Poul Anderson once said, 'I have never seen any situation, no matter how complicated, which if you look at it in the right way, can't be made even more complicated'. My approach is the opposite; as a theoretical physicist I say, 'Of course the world is always complicated.' But, for many things, if you look at it in the right way there are underlying simplicities that are really important to understand what's happening – and some things that are less important. And you use a mixture of intuition and guesswork to pick what might be the important things, and then – and this is the critical thing – then you express that intuition in a very precise way, which usually means in a mathematical way. So I see mathematics as no more, but no less, than a way of expressing thoughts clearly.

Then if you've expressed it mathematically, you can draw conclusions, and you can circle back and see if the conclusions – sometimes about things you haven't looked at yet – turn out right; and if they don't, you ask what's different. So, it's a way of using mathematics as a tool in trying to understand the world. It's only one among many tools that experimentalists will want to use, and it's a way that I happen to have some skills at. And, if the truth be told, personality and style play a part in the way we do things.

I just enjoy puzzles, and in a sense that's a way of going about it; in some ways it's not all that different from looking at a chess game.

ah-d

But the maths can be hideously complicated. Do you find it difficult to explain to other scientists why this way of looking at it is important, or even to explain to them how to look at it in this mathematical way?

rm

Well I'll answer that in two ways. First of all I'd say it's always easier to explain something that's purely descriptive than something that has been abstracted into a mathematical equation. But if you really understand what is going on, and if you're willing to make the effort, I think you can convey it. People as varied as Martin Gardner or Ian Stewart can explain immensely complicated mathematical things in ways that can be generally understood. It's a matter of making the effort and ... and caring about it.

There's a second problem, which is this: there are hugely varied styles of doing science, and people who spend their life in research bring to it a hugely varied range of skills. Some of them are people who from an early age are comfortable with mathematical things, and the others are people who like to take things apart and play with them. Some of the mathematical people couldn't fix the washer on the tap, and some of the people who can do wonderful things with bits of machinery or build robots for robot wars and so on just aren't very good at mathematics.

For some problems you need all those skills, and people can be combined in teams. I think it is true that some newly developing subjects go through an early stage that is purely descriptive, but as they mature they reach a stage where a degree of theoretical structure becomes helpful. At that point many of the people who were inhabitants of the purely descriptive stage will feel uneasy, not to say resentful. And others will be excessively enthusiastic.

So in my life, having been trained in the fully mature discipline of theoretical physics, where nobody in a physics department ever disputes that you should be have experimentalists and theoreticians, I found that, in the early seventies, in ecology, there were some people who felt these people writing down equations about animals were just a bunch of ponces mucking around. Because they believed that to really understand, you've got to get up there with muddy boots and suffer in the cold and realize it's agonizingly complicated, and the more you look at it, the more complicated it is.

Conversely, there were people who were excessively uncritically enthusiastic and there have been some rather silly things done. And there were people who came in with a box of tools who wanted to find a problem they could apply their mathematical tricks to, which invites a certain amount of irritation. Nowadays any contemporary ecology text will have a blend of theory, observation and experiment, as the subject has matured.

If you look at immunology, there are still some immunologists who are doing utterly brilliant things at a molecular level, who are a bit distrustful that looking at mathematical models of populations of viruses interacting with populations of cells has anything to do with what they're doing. That too will gradually change as the subject matures.

ah-d

You spent five years as Chief Scientific Adviser to the British government. What's it like trying to talk science to ministers?

rm

Well I never actually did much mathematics with the government, although I did publish one neat little paper with one of my colleagues on queuing theory, pointing out that most people in Whitehall thought that if the rate of sending papers to a minister went up and the minister was slower to move them across his desk, it would just mean that things went a bit more slowly. We pointed out that there was a critical point in this process where if the clearance rate went significantly below the input rate; it wasn't that things slowed down, but that the whole system collapsed. I wrote a little note on that, which caused some amusement and even had some impact.

The nature of the job of the chief scientist is first of all to try to maintain an oversight, and make sure that all the departments are doing the things – mainly routine things – that they ought to be doing in a competent way, which they nearly always are. The really exciting things, which crop up relatively rarely but often with huge impact and get all the attention in the media, are those beyond the frontiers of what we understand – whether it's BSE or how best to handle Foot and Mouth, which you might say we ought to know because we've had it before. But farming practices have changed, and the web of connections among farms is different today.

Then there's the advent of GM foods, where people ask questions that need to be asked. Are there worries about food safety? Are there worries about the impact on the environment? In cases like those you want to make sure that you get the very best people and that you listen to all the voices and that things are done as openly as possible. I think that's a very interesting challenge because the world itself is changing. The job of science in government isn't the same as the job that's always been done.

Thirty years ago we lived in a world much more respectful of authority and you could get away with confidential advice to ministers, and ministers saying, 'The experts told me that.' Thank God we live in a much more open world today and people don't want to be told – they want to say to the minister, 'What's the advice? What's the basis of the decision? Let's see the process. Let's join in the process.' I found it an exciting time to be in government as that old cake of custom was being cut and the notions of confidentiality and cosy coteries of advisers were being subjected to a much more open process.

ah-d

Did you have power? Take the GM foods, for example. Did you have power to go and ask people for information?

rm

Well I would say the word 'power' is not one that occurred to me. The word 'influence' is what you have – power, no. With hardly a single exception, if I wanted to ask someone's help in something, whether they were outside government or inside government, I got it. The problems were of trying to reconcile different approaches and different opinions or, in the case of GM foods, of trying to conduct, in a policy arena, the kind of discussion that's so common in science, where you get half a dozen different opinions, all fiercely held, all talking at once. And that's something that serves science well, but is more awkward in public policy.

I mean, someone produces an idea, someone else produces another idea. The ideas contend, the arguments are heard and, in the early days, when we didn't know so much, there develops a landscape of different opinions – one group of opinions here, another here. Gradually over time we learn more until, finally, there's an agreed understanding, and now you can set it in an exam paper and know what the right answer is.

Science is about recognizing and acknowledging uncertainty.

Government's much less accustomed to such uncertainties. Most people experience science in primary school, secondary school or in university, or on quiz shows – *Who wants to be a Millionaire?* or *The Weakest Link* – as certitude. That's understandable because you're going to teach a curriculum of things that we really know about, and you've got to have a right answer and a wrong answer on a quiz show.

But many of the problems in today's real world, and in particular many of the problems that come across the desk of a chief scientist, are things beyond the frontier. That's comfortable to a research scientist, striving towards a better understanding and knowing you're not certain. But it's very uncomfortable to people who think that science is certainty. It's understandable they think that, but it's unfortunate. Because at the frontier, science is about recognizing and acknowledging uncertainty.

Take BSE: the first question is, 'Is this mad cow disease in cattle going to infect people?' The answer back then was, 'We don't know.' Overwhelmingly the best guess was no, by analogy with scrapie in sheep. Scrapie is like BSE, and to the best of our knowledge it has no effect on people. So what we should have said – this was all before my time as chief scientist – what one would like in retrospect to have said, and very clearly, is, 'We don't know.' But that's an uncomfortable thing to say. 'Here's our best guess. Here's the reason for our guess. Now you decide.' You can understand how people might then have said, 'Oh my God, that will do damage to the beef industry and it's almost surely OK,' and it quickly slips into the minister saying, 'There's nothing to worry about, and here's my daughter eating a hamburger.' What we are trying to do now is create a different world in which we share the uncertainty; we give the basis for the guess, and leave it to individuals to make the choice. This isn't easy, but I believe we have to learn to do it.

ah-d
We've got ourselves into a bit of a pickle with genetically modified food, haven't we?

rm
I would say there are three categories of worry. One is whether there are novel worries about the safety of the foods themselves. Then I'd distinguish two kinds of environmental worries. One is worry whether they could cross-hybridize with other things and create superweeds. And then there's the more general worry of further intensification of agriculture. And my own personal view is that we should be looking critically at all novel foods, because food is dangerous stuff – there are many kinds of conventional foods that can do nasty things if you're not careful, and if they are not cooked properly.

So we should always make sure we're not doing silly things with any foods, whether GM or produced by so-called conventional breeding – which, if you knew enough about some of the recent techniques, can look every bit as weird as specifically targeted gene insertions. We know that many of our currently conventionally bred crops can, to a degree, cross-breed with other things and create hybrids. And we need to worry: are we going to produce superweeds? I don't worry too much about that because, by and large, our crops are the last place you would look to create weeds. There are still questions to be asked, but basically crops have been bred to produce food for us, and in breeding them we've generally made them less able to survive in the wild.

My real worry is a much more difficult one, which is, from the dawn of agriculture, our dream has been to produce crops that only we eat – that we don't share with weeds (which are just plants in the wrong place) – and that we don't share with insect pests (which are just insects with the wrong appetite). And the closer we get to realizing that dream, the less there is for the rest of the natural world, and the more silent spring becomes. I welcome the debate on GM foods, even though I think some of the worries about food safety are distractions. I welcome it because I welcome a chance to ask much clearer questions about what kind of countryside do we want – what kind of world do we want. And how do we reconcile really efficient agriculture with the interests of wildlife? You've got to feed a lot of people, while at the same time trying to preserve biological diversity.

What kinds of use do we want to make of these new GM techniques? The first phase of genetic modification has been driven by the interests of agribusiness, and it hasn't produced benefits for the consumer. Which is why it's quite sensible that people in the UK say, 'We hear these worries, Greenpeace says what it would say, the government says what it would say. But in the meanwhile there's nothing we want to buy. So let's have none of it.' In the second wave I believe there are going to be products that serve the consumer.

In the developing world we will, I hope, use these techniques to produce plants that grow with the grain of nature – where the plants fit the environment instead of the environment being wrenched with fossil-fuel energy subsidies to fit the plants. Plants that are drought-tolerant, salt-tolerant. Maybe ultimately plants that fix their own nitrogen. And even in the developed world we may learn to produce allergy-free nuts, or ultimately a golden apple that you can eat to become thin and witty.

To me it's a debate that is valuable for itself, but even more valuable because it's a rehearsal for yet more complicated and difficult debates we're going to have over the coming decades as our increasing understanding of the molecular machinery of life gives us the ability to do things that we should ask deep questions about before we do them.

ah-d
You're now President of the Royal Society. Presumably you have no time for research any more? And do you think the gulf between the public understanding of science and real science is getting worse?

rm
Well, let me say two preliminary things. The Royal Society sounds like something that produces the Queen's address book or something, but in practice the Royal Society is the UK analogue of the US National Academy of Sciences. It's the United Kingdom and Commonwealth Academy of Science. It was created by Charles II and called the Royal Society because it happened to be the first society to be given a royal charter – but it's a rather kinky name.

Secondly, I do still do research. I have a research group. During the five years I was chief scientist I spent a day a week in Oxford, and I actually wrote about a third of a book with a colleague, Martin Novak, on virus dynamics. And I've got half a dozen refereed papers to my credit just from the last year. I now spend half the week in Oxford. But the other half of my time I spend on affairs of the Royal Society.

Do I think people distrust science more? Well, the motto of the Royal Society is a cryptic one, in Latin, *nullius in verba*, which is taken to mean 'Look at the facts'. The facts about public perception of science come from a variety of public opinion polls, which suggest that 84 per cent of people say 'Science has made our life better'. And two-thirds of them say,

'The aim of scientists is to make life better'. That's reassuring, and similarly university professors are generally rated as 'good people'. Government scientists don't do so well.

By and large people think science is on the whole a good thing, but – here comes a big but – roughly half the people in the same polls say, 'The pace of modern scientific advance is too fast for the government to be able to keep up with effective regulation'. And again, frankly, I share those views. I think the public has a great deal of common sense. And I share some of their worries, which deserve attention.

The notion, however, that people today distrust 'the new' more than they ever did is exactly upside down. You go back a few centuries and the bringers of the new were tied to a stake and burnt – they burnt Giordano Bruno for suggesting that the sun didn't go round the earth, although Galileo got away with house arrest. Two hundred years ago at the advent of smallpox vaccination, there were riots in the street, cartoons in the broadsheets, fully in the contemporary European idiom of Frankenstein food cartoons. A hundred and fifty years ago people suggested that if you got on a train and went too fast through the long Box Tunnel near Bath you'd get asphyxiated.

ah-d

So do you think things are getting better?

rm

I'm saying I think most people think science aims to do good and has made their life better. I personally think science certainly has made our lives better, no question about it – the best of times. On the other hand, it's had unintended adverse consequences that no one foresaw, such as climate change and diminishing biological diversity, and we have to wrestle with that – the worst of times.

chapter NINE

John Maynard Smith

Why we bother with sex

John Maynard Smith was born in London on 6 January 1920. He graduated in engineering at Cambridge University and worked as an aircraft engineer throughout the Second World War, calculating the stresses in aeroplane wings. He then took a second degree, in zoology, at University College London, and became a lecturer there. He moved to Sussex University in 1965, where he was Professor of Biology and later Professor Emeritus. He has applied his knowledge of mathematics to evolutionary problems, using game theory to explain the paradoxical cooperation of male combatants in many species. He has tried to apply the theory of natural selection to the design of robots and computers, and also worked with a group of neurobiologists on the behaviour of ants, bees, worms and snails in what he called "The Institute for the Study of Tiny Minds". He died in April 2004, soon after approving the text of this chapter.

ah-d

Professor John Maynard Smith is one of the grandest of biology's grand old men, a man who put game theory into biology – and who has spent most of his long life thinking about sex. John, why do we bother with sex?

jms

If I knew the answer to that question, I would be happy. Sex is a puzzling problem for an evolutionary biologist. I think the main contribution I've made is just to say, 'Oi, look, this is something we don't understand and ought to think about.' The orthodox answer, which is part of the truth though not, I think, the whole truth, is that a species that has sex can evolve much faster to meet changing circumstances, because genes that arise in different ancestors by mutation can join together in a single descendant. Whereas if there were no sex, there would be no way these good genes from different lineages could ever get together. If you do the sums, it turns out that species do in fact evolve much faster if they have sex.

So sexual reproduction is good for the species. But that doesn't really answer the question whether in the short term it would be good for the individual female to abandon it. You might say that we bother with it because it's fun – and it is fun, I suppose, because things are fun for us if they're going to increase our Darwinian fitness, if they are successful. One can also see why males bother with sex. They have no other way of passing on their genes. But why on earth should females bother with it? They would do so much better if they produced female offspring like themselves by virgin birth. They would in fact have twice as many descendants like themselves as they do now.

I honestly don't think the amount of parental care that the female gets out of the father is worth bothering with. I think she'd do a lot better without it. And yet to abandon sex is unusual. There are organisms that do without sex, but not too many: bees have cloned sons, but they have to have sex to have daughters, which seems funny.

ah-d
Is there a lot of cooperation between individuals?

jms
There has, in a sense, to be cooperation between a male and a female in a sexual species in order to produce an offspring. But it's not conspicuously cooperative, except in that they both have something to gain, i.e. producing offspring.

If you do the sums, it turns out that species do in fact evolve much faster if they have sex.

ah-d
We're told by the likes of Richard Dawkins that genes are selfish and that the 'aim' of each gene is simply to reproduce itself, but you're saying that there has to be something for the advantage of the species; not only do the genes on a single string of DNA have to cooperate, but there has to be cooperation all the way through the body and indeed through the kingdom. How does that work?

jms
I think you've put your finger on what may be the hardest question in biology, exactly why sex is a puzzle. It's certainly a question that a lot of people are thinking about very hard right now. The orthodox explanation of it depends not on the advantage of the individual gene, but on the advantage of the population as a whole. And this makes both Richard Dawkins and myself, I think, very uncomfortable. So we look for the shorter-term advantages for sex, if we can think of them.

There are some simple answers that explain some of the things. If you've got eight men in a boat, each with one oar, it really wouldn't pay one of them to stop rowing. If he wants to get somewhere, he's got to row; he can't get out of the boat and swim. And I do think that of the genes on a chromosome: at first you say, 'They've got to cooperate, otherwise how will they ever get passed on?' But although that's partly true, it's not the whole truth, because genes don't cooperate; they jump from one chromosome to another – jumping genes. The more we know about the details of what genes are doing, the harder it becomes to explain.

We're a terribly indoctrinatable species. We're happy to believe any lie people tell us when we're young, and we finish up believing all sorts of totally crazy things. There are reasons why human beings have these properties of being teachable, and also why societies have policemen. The question is, can we extend that back to animals? Is there an equivalent of the police in the animal world? And I think, yes there is. Cheating is punished in the animal world, to some extent – and yet human beings are very often beastly to one another. That's just part of the pattern of conflict in evolution between two processes. There is a process that tends to make genes or cells selfish – they become cancerous – or to make individual people selfish – they become criminals – or to make a group selfish – they become terrorists. Even species can be selfish; at each level one can find selfishness, and yet one also finds cooperation, and there's a constant tension between these two processes. I think that's what a lot of evolutionary biologists who are really interested in the deep theory are thinking about right now.

ah-d

Another thing that always seems a bit odd to me is that we have roughly equal numbers of boys and girls. Is it obvious why that's the case?

jms

I don't know if it's obvious, but I think we do understand that. I can explain it like this: supposing you were a female, a woman, and you were able to choose the sex of each child, boy or girl – which sex should you choose? An evolutionary biologist would say, 'You must choose whichever sex will enable you to pass on your genes to the greatest number of grandchildren'. If there are fewer boys in the population than girls, boys will have on average more children – and vice versa. So if you're able to choose the sex of your child, you should choose whichever is the rarer sex. The only stable state of the population is then to have equal numbers of both. The XY chromosome segregation is just a way of bringing that about.

On the other hand, I'm sure that if it were advantageous to produce, for instance, three daughters to one son, natural selection would have found a way of doing that. The one-to-one segregation is just a way of ensuring that there are equal numbers.

ah-d

I'm told you are the biologist who put mathematics into biology. Is that right?

jms

Well, I wouldn't want to claim that. I think my teacher, JBS Haldane, put mathematics into biology and taught me how to do it; I've been following him ever since. But certainly I have spent my life using mathematics in biology.

The argument we've just talked about, for a one-to-one sexual ratio, is something that requires mathematics. I don't claim that I was responsible for that. I think it goes back originally to a man called RA Fisher. But the idea of looking for a stable strategy is one that I came up with and I'm quite proud of.

ah-d

Is that to do with game theory?

jms

It is indeed. In fact, I suppose the thing I'm best known for is introducing game theory into biology. Game theory was originally invented to talk about human conflicts, anything from playing poker to gambling in the City. The idea was, what should people do if they're rational? What I did was to say, let's take the same mathematics but ask what should animals do, not if they're rational, but if they behave in such a way as to maximize the number of offspring they produce. This is still essentially game theory.

I first of all draw up what is called a pay-off matrix, which is an idea I borrowed from the game theorists. I can say the animal can do this or it could do that. It can fight with all its weapons or it can display and run away if the other chap attacks. If we call the two guys hawks and doves, what is a hawk? What does a hawk get, on average, if its opponent is a hawk? What does a hawk get, on average, if its opponent is a dove? And the same for the doves. Then you just do a little sum and find out what will happen.

ah-d

Does that really help in explaining behaviour?

jms

I think it really does, yes. I'm not sure I thought so when I came up with the idea in the first instance. But I can remember vividly that I predicted, from this very simple model of hawks and doves, that you would expect to find animals having what I call respect for ownership. If I'm the owner of something and you're an intruder, I behave like a hawk, and you behave like a dove and run away.

At first I thought that animals can't possibly have respect for ownership; it's absurd. But then I went down to Texas – I was in America at the time – and gave a seminar about the game theory that I'd just invented, biological game theory. And I finished up by saying, 'So you'd expect animals to respect ownership – well, that's what the theory says – but of course I don't really believe it.' Then a young man whom I've loved ever since got up and said, 'Can I describe to you the work that I did for my doctorate in California?' And he told us how he worked on butterflies, how the males occupy hilltops, and a female who wants to mate flies uphill till she gets to the top and then mates with the male. The only snag is that there are more male butterflies than there are hilltops, so the males get involved in fights about hilltops.

What this young man, Larry Gilbert, had been able to show is that the owner of the hilltop fights like a hawk, and the intruder behaves like a dove. I remember sitting there thinking, gosh, I was right after all! A great moment. Some animal has read my papers, and is doing the right thing.

As another example of such a moment, many years ago I wrote a book about game theory, and I pointed out that in principle you can have a game that doesn't have a stable point – that will cycle forever. The game I mentioned was a kid's game, Rock, Scissors, Paper. The trouble is that rock beats scissors, scissors beat paper, paper beats rock. So these strategies would replace one another forever in a cycle in the population. I mentioned it as a purely theoretical possibility – and then I opened *Nature* one day quite recently, and there was a paper entitled 'Lizards play Rock, Scissors, Paper'. This particular species of lizard does – at least, it doesn't do exactly this, but it has three mating strategies, and they cycle in the population, one after the other. I felt really pleased.

ah-d

Going back to the insects, you once did some curious experiments with insect flight, did you not?

jms

I did work on insect flight, actually in order to prevent myself going mad after the war. I'd spent the war designing aeroplanes, and then I went back to take an undergraduate degree in zoology. I decided I didn't want to spend my life designing aeroplanes, I wanted to become a biologist. But I was 28 or so, and I found just learning and listening to lectures was driving me mad. I had to do some research. So I did some research on animal flight. Then I studied how insects fly, which was great fun.

I think I was the first person to take a photograph, not of the movements of the wings, which are too fast, but of the flow of the air around the insect. You have your insect and you glue the head of a pin to its thorax; you can hold it there, and it will beat its wings – if it's the right insect. These were big flies. You know that the air must be being sucked down by these wings and chucked down in a jet. The problem is that you can't photograph air either, so we invented an instrument new to science called a flufficator. It filled the air with fluff – anybody who likes designing things, you need a baby's sieve, a block of metaldehyde and a soldering iron, and you too can make a flufficator. We surrounded the insect with fluff, and we took a photograph with a flash bulb. We got lovely lines on the photographic plate, giving the direction and velocity of the wind. We also measured the velocity – that involved a wind-up gramophone. I had a reliable son, I think he must have been about ten at the time, who would keep winding the gramophone, which was used as a clock.

ah-d

So you started life as an engineer, you were an engineer through the war, and then you switched into biology. How did that come about?

jms

I think I'd always wanted to be a biologist, but when I was 19 the war broke out. Before I went to university, I simply didn't know you could earn your living being a biologist. It had never occurred to me. And so I went to be an engineer because my mathematics would be useful there. When the war ended, I realized that you could earn your living as a biologist. I went straight to Sussex University, and I was the first biologist there; I had to invent the biology department. Very hard work, but lovely, nobody to control me at all. I could do what I damn well pleased.

The genome is full of lovely information, but most of it is illegible at the moment.

ah-d

There is a famous book by Schrödinger called *What is Life?* which must have been published around the time you were taking that decision. Was that important?

jms

Not to me personally. But I think it was important in a more general sense. Schrödinger was a theoretical physicist; I didn't meet him till many years later, but there's no question that he did influence a number of very bright young physicists to get into biology. He was really saying, 'Look, boys, this is where the action's going to be'. And they were marvellous;

they were the fathers and mothers of the molecular revolution; they've completely changed everything.

ah-d

In those days it seemed that physics was the dominant science, that it was going to take over the world and run everything. Nowadays it seems to have swung round; you biologists are the scientists who claim everything. But now that the genome has been worked out, is that the end of the road?

jms

No, because we can't read the thing. The genome by itself tells us very little. It's rather as if we were presented with a Hungarian dictionary but didn't speak a word of Hungarian; it wouldn't enable us to translate Hungarian books. The genome is full of lovely information, but most of it is illegible at the moment. We don't know what it means.

ah-d

You yourself have recently written a book about the origin of life. Is this something that continually exercises you?

jms

Yes indeed, I think every evolutionary biologist has to worry about how life began. But one also has to worry about whether it's really possible to go from the first simple replicating molecules in a primeval soup to things like ourselves or oak trees. Have we really got a theory that will explain that? That is what this book, which I wrote with a young Hungarian called Szathmáry, is trying to answer. I'm sure we can explain it – we do explain quite a lot of hard things. You asked me earlier about one of the things that I think we do not explain very well, which is the origin of sex – I think people are puzzled by that. But many other things we do explain.

I do think that we actually could start with non-living matter and see the origin of life, and that we could then claim to have understood how it happened. I wouldn't be surprised if someone does that within 20 years. We have a pretty clear idea of how most of the steps took place. We talk in our book, for example, about the origin of what's called the genetic code: the process whereby a replicating nucleic acid molecule can evolve the ability to specify proteins, and how the proteins will help to replicate the molecule. We have, I think, a rather convincing theory of how that came about. We'd all buy the origin of cells and of multi-cellular organisms and so on, and most of those transitions we can have a shot at.

The trickiest one is the great transition that led from other animals to humans, which is a quite special event. I think that we are pretty clear, or I'm fairly clear, that the crucial change was the origin of language. I think essentially we're different because we can talk; we can transmit information between individuals and between generations – not just by waiting for our genes to change, but by speech.

I'm willing to predict that if we were to find intelligent life on some other planet, it would certainly talk in some way. It would have some linguistic skill for conveying information as you and I are now doing. That has to be so – otherwise there is no way it could achieve the kind of things that we've achieved technically. Intelligent advance requires intelligence. You have to be able to think in order to make complicated things – or even fairly simple things.

ah-d

Are you saying that we can't think without language?

jms

Some people claim that we can't, but I suspect they've never played chess. Clearly we can think without language. But there would be no history, no continuous change in society, without language. And if we didn't have history, we would remain a rather simple mobile ape wandering about the Savannah killing things.

ah-d
You're talking as though information is a crucial part of life. I thought living things were just things that ate and drank and died.

jms
If you read a book or you listen to talk among molecular biologists, their talk is full of informational terms. There's translation and transcription and editing and proofreading and libraries – all technical terms in molecular biology. It is quite clear that molecular biologists think of the whole genetic mechanism as a machine for transmitting information from one generation to the other, and I think they're entirely right.

ah-d
And you say information transfer is now much faster. Passing information on by DNA takes 20 years, but by word of mouth it takes 20 seconds.

jms
That's why history is quite different from evolution. History has transformed our technical abilities in ten thousand years. It would take ten million years to make such a major change in evolutionary time.

ah-d
Are you saying that we now have a second replicator?

jms
We have. This also makes me think of major transitions in human history. One can claim that the origin of human history was language. But before one can have large societies, one has to have not merely language but writing. You

can't tax people if you haven't got writing; you can't build a complicated civilization. The next major transition in history was the invention of printing. And we're now living through the final one, namely the electronic transition. Of course we have no idea where that will end – I haven't anyway.

ah-d

So we've got genes, which replicate once a generation, and we've got these ideas, these memes, which replicate in microseconds. Do you go along with the idea of memes?

jms

Yes, I quite like the idea of memes. A meme is simply a replicating idea. The simplest example of it would be something like a limerick – 'There was an old man of Japan …' I tell you that, and if it's a good limerick you will then remember it, and tell somebody else, and he will tell three or four people and so on. The thing will spread like a successful gene.

The parallels between the genetic language and the human language are really quite extraordinary.

ah-d

So the memes drive us to have language, because language allows us to digitize information.

jms

Yes, and it's interesting you use the word digitize, because the parallels between the genetic language and the human language are really quite extraordinary. Our human language starts off with a rather small number of discrete sounds, some thirty or so, B and P and all the things we symbolize by the alphabet. Then we fit them together in words, and then we fit the words together in sentences, and the genes do just the same. They have an alphabet of only four primitive sounds, if you like, namely the four bases, and they fit those together in words which are called genes. Then they string genes together into little sets of instructions like sentences. The analogy of human language and genetic language is extraordinarily close. It sounds simple – but I think big ideas usually are simple.

ah-d

Is there an advantage in coming into biology tangentially as you have?

jms

Yes, I think there's an advantage in any science, but particularly in biology. The thing that really helps you to make original discoveries is what you know that other people in the field tend not to. In my case, when I started, that was mathematics. I wasn't the only person with mathematics, but I was one of only a few. Even bigger changes were made to biology by people who came into biology knowing some serious chemistry – they were responsible for the whole molecular revolution.

I think that good evolutionary biology is done by putting together the detailed odd facts of natural history, and then putting those together with some more abstract theory, just as Darwin did with the finches – and we're all following Darwin now.

So I think that it did help having been an engineer. It taught me mathematics, and what is more it taught me how to use mathematics: in my case how to make models of the aeroplane, models of the world in mathematical language.

ah-d

Forgive my saying so, but you're older than I am, and by more than a couple of years. And yet you're still going into work every day. Isn't that a bit masochistic?

jms
No, I only do it because I enjoy it. I mean, they don't pay me!

ah-d
You do it for fun! I thought science was supposed to be difficult.

jms
Chess is difficult too, but people play chess for fun.

ah-d
That's true. Do you have a great research team?

jms
No, I tend to work on my own, except that I like to be surrounded by a few young people who I can talk to, and who will help me with certain skills that I haven't got. I'm no good at getting into the database through a computer and collecting DNA sequences and so on. My young friends do that for me. And they give me ideas that I can steal.

ah-d
What about communication? Are you a wizard on the email and the Internet and so on?

jms
No, I intend to be the last human being left alive who is not on email. And as for the Internet, it's a total mystery to me. But unlike most people who play with computers, I'm a very good programmer. A lot of my research is done by writing really quite complex programmes through a computer. But nobody seems to do that nowadays.

ah-d
No, it's all disappeared. So what are you working on at the moment?

jms

I'm working partly on sex – but sex, oddly enough, in bacteria. I've become very interested in the way that bacteria share genes, swap genes, how genes from different bacteria come together in a single descendant and so on. It is also of very great practical importance; it's been extremely important in the evolution of antibiotic resistance in bacteria that is such a worry at the moment.

ah-d

Isn't there a theory that one reason we bother with sex is to escape from the parasites?

jms

Yes, the idea goes back to Bill Hamilton, who suggested that whatever the ultimate explanation for sex, an obvious one is that it helps us evolve very fast. The objection usually was, 'But why do we have to evolve very fast? Temperature and climate and so on aren't changing all that quickly; why should we bother?' His answer was, 'Yes, but the parasites are evolving very fast.' Diseases are our viruses, our bacteria; we have to be able to evolve rapidly in order to cope with our evolving diseases, and they have to evolve in order to cope with us.

ah-d

It sounds like your game theory all over again. Is there a solution to the problem that the common cold is changing all the time, and that simple viruses are becoming resistant to antibiotics?

jms

It's a very worrying problem. I think we have behaved madly in relation to antibiotics. The idea of putting antibiotics into animal feed, and of doctors prescribing antibiotics for the common cold when they know perfectly well that it won't do any good, just to get rid of the patient – these things are criminal. The maddening thing here is that geneticists knew this perfectly well 30 years ago. Not me, but others, and they said so. As far as I can see, absolutely no notice was taken.

ah-d
Most of the time we tell the truth, but obviously there are some situations where it's worth lying. To the biologist, do people lie a lot?

jms
I'm sure all of us lie sometimes. A major reason why we do not lie more is that if I tell you lies, you will remember it and say, 'I don't want to trust that chap, I'm not going to believe him tomorrow.' And not only that; you'll also tell your friends, 'Don't accept a word of what Maynard Smith says; he's a liar.' I would get a reputation as a liar, and that would do me no good at all. I think that does explain why human beings are generally honest. It doesn't, however, explain why animal signals are honest, and that's a much harder question.

ah-d
You mean, animals have the option of telling lies?

jms
Yes, there are many animal signals. For example, sparrows vary in the size of the black bib they have. The male with the large black bib is signalling, 'I am very aggressive and if you tangle with me, I'll attack you', whereas the male with a much smaller black bib is saying, 'I'm not really a very aggressive chap, leave me alone, I'm a friendly sort of chap'. Now if, very unkindly, you take one of the latter, the doves as I call them, and you paint him with a bigger black bib, he's lying. He's saying, 'I'm going to be aggressive', but actually he runs away. He has a really rough time;

he is driven out of the flock. You don't want to do it unless you can be quite sure you can catch him and wash him clean, otherwise you're really sentencing him to death. The others don't recognize him as an individual, but they are wired up to know that this chap's lying.

So animals don't lie very much. There are always some lies, of the sort we call mimicry: there are orchids that pretend to be bees so as to get fertilized. There are lots of mimicry lies, but they depend upon there being an honest signaller out there, in order to mimic the honest signal.

ah-d
Where will evolutionary biology go in the next 20 years?

jms
I think if you want to know the future, you should ask a writer of science fiction rather than a biologist. I think HG Wells was a much better predictor than his contemporary scientists; a man called Olaf Stapledon was a marvellous predictor who wrote science fiction books that I read when I was 16 and that completely blew my mind; and Arthur C Clarke put his finger on quite a number of bright thoughts. He and I have something in common: we both took out of the public library the same science fiction book when we were boys of about 15 or 16, which was Stapledon's *Last and First Men*. We took it out of the same country library in Porlock in Somerset. Whoever put that book on the shelves had a lot to answer for!

I don't know what's going to happen, but there aren't too many good science fiction writers around either, so perhaps you'll just have to guess for yourself.

1 Strap onto your LEFT hand.
2 Turn dial toward fingers until it STOPS.
3 Relax for 5 minutes.
4 Turn dial away from finger until light is DIM.

*chapter***TEN**

Rosalind Picard

Affective computing

Rosalind Picard was born on 17 May 1962 in Boston, Massachusetts, and graduated in electrical engineering at the Georgia Institute of Technology. She worked for AT&T Bell Laboratories on digital signal processing and methods of image compression and analysis while doing postgraduate study in electrical engineering and computer science at the Massachusetts Institute of Technology. She subsequently joined the Media Laboratory at MIT. Her current focus of research explores areas such as the sensing and synthesizing of emotion in machines. Her book *Affective Computing* (1997) lays the groundwork for giving machines the skills of emotional intelligence; its thesis is that emotion is necessary not only for humans to think rationally but also for computers to be truly intelligent and to interact naturally with their users.

ah-d

Professor Rosalind Picard is a lady who likes to wear a computer on her sleeve. You've written a book, Rosalind, called *Affective Computing*. What does affective computing mean?

rp

Affective computing is computing that relates to, arises from, or deliberately influences emotion. It's about giving computers some of the skills of emotional intelligence, like the ability to see if they've irritated you or not. I'm not so sure you'd want computers to get angry back at you, or to have feelings like people do. In fact, it's even scientifically questionable whether they can have feelings like we have. But they certainly can give the appearance of having feelings. In fact, they already do in some ways. In science fiction we've seen for years how the writers of these stories, such as *2001: A Space Odyssey*, have portrayed computers with feelings, and it's easy, for example, to have the little paper clip on the computer smile at you or wink at you.

ah-d

So you don't want to give computers feelings, but you want them to be able to recognize our emotions, is that right?

rp

I am working primarily on having them recognize our emotions, but it's also possible that we might need to give them something like human feelings in order for them to respond intelligently to what they've sensed from us.

ah-d

Go through the first stage. How are they going to recognize our emotions?

rp

Well that's a research question. There's no simple signal that allows us to tap into your body and extract what you're feeling. In fact, most people can't even say accurately what they're feeling. So our first measure is how well people can do this. Given that, we start to break down how they do it. For example, people can read facial expressions. You can smile and frown, but if you're smiling it doesn't necessarily mean you're happy. There's the fake smile versus the real thing.

The computer can look at your face with a camera, it can track where your eyes are, and where wrinkles and lips and different features of the face are. It can watch the movement in time, and then say, this motion pattern looks like a smile. Then maybe when it sees that smile, and it also sees you, or hears you laughing or sees you leaning forward, sees you behaving as if you're enjoying yourself, it might infer that this user seems to be having fun.

ah-d

But a smile is far from trivial, isn't it? As you said, just the shape of the mouth is not necessarily relevant. There's a whole lot of other stuff.

rp

That's right. For example, if it's a genuine smile, the eye muscles move too. When you're faking a smile, those muscles don't move. This is why photographers know to try to make you laugh or tell you something funny to get a genuine smile.

ah-d

Isn't it true that Marvin Minsky once gave a graduate student the task of sorting out computer vision?

rp

Yes, it was a summer project to see if he could write a piece of software to tell a dog from a cat, by sight. And about fifteen years later, somebody succeeded. They've been working on it for decades. Computer vision is a funny problem; it's in the camp of problems that look easy. You just open your eyes and you see; what's hard about that? But it's actually very hard to get computers to do it. It's the bizarre little things that throw them off. You can have the computer recognizing your smile and your frown and your eyebrows, and then as soon as the lights change, suddenly the computer can't even figure out where your face is.

We're amazingly able to pick out what's important in what we're looking at and to ignore all the irrelevant stuff. But we don't even know how to give computers the kind of attention shifts that we people have. We think that the emotions inside you are largely responsible for regulating things like what we decide to direct our attention to.

Because vision is difficult, we are also trying other methods. For example, we decided that if you're wearing your computer, then your computer actually has an opportunity to sense some things more easily than facial expressions. We're now able to sew computers into your clothing, in a way that looks more like clothing than like computation. There's a new conductive thread that can function like electrodes, or carry regular circuit information.

We've developed a sensor that is wearable. It's skin conductivity-sensitive; there's a tiny little board of electronics inside a pouch, with a replaceable battery and a little dial and an LED. Actually, it was a cycling glove that inspired this. The light is proportional to the amount of skin conductivity being sensed across two electrodes on the palm. The electrodes are just regular clothing snaps that we've hooked wires around and punched into the fabric.

Usually, skin conductivity is measured across two fingers, and we tried that originally. But we wanted to be able to type on our computers, and the electrodes got in the way; so we decided to move it to a more convenient place on the palm of the hand. This turned out to be a nice position, because you can actually do most of your daily activities wearing the sensor, and watch how your skin conductivity changes all day long, from when you wake up in the morning and your light goes dim, as you're really groggy, to when you've just had that first cup of coffee.

It does seem that your adrenaline level affects the conductivity. In fact, at the end of the day it can be very hard to get yourself to glow. This is just a basic physiological measure. It's related to the amount of perspiration on the hand, so if you exercise, the sweat will make it glow. In fact, sometimes even when your palms feel dry, the signal can change significantly due to very small changes in the sweat glands that might be triggered even by something such as thinking an embarrassing thought, or thinking of something that you were supposed to be doing right now, so that you have a little feeling of urgency or of significance. These little feelings are physiological signals in your body, and we can tap in and measure them through this signal.

ah-d

So, this is more or less like a lie detector, isn't it?

rp

The skin conductivity that this measures is one of the signals that has traditionally been used in lie-detector tests. However, it's not reliable by itself as a detector of lies. It can, for example, be fooled by things such as sticking a thumb-tack under your big toe and, every time they ask you a question, stepping on it a little bit. Pain is significant, and overrides it. Similarly, if you think of something that embarrasses you, you will have a little feeling that you can notice, and that will also show up in a change in the signal.

This alone doesn't tell the computer whether you like it or not, or whether you're confused or frustrated. However, it is one indication among many that can help present a bit of a picture of what the user is feeling right now. For example, we've combined this signal with three other signals, and we've taught the computer how to extract patterns from those signals, and how to determine which of eight emotional states a person is experiencing, including neutral, grief, anger, joy and two flavours of love – platonic love and romantic love cause different physiological patterns of response.

The computer can then learn what your physiological patterns look like as you're experiencing these different states. It can then watch you and say, oh yeah, he's in his neutral state right now; or, oh dear, he's looking like he's getting agitated. Maybe it doesn't have anything to do with what the computer's done, maybe it does. The computer can take a note of what it was just doing with you, and see whether there's a pattern. Maybe the computer has noticed that you get agitated every time you see that little paper clip, and it could offer to turn it off for you, completely disable it so you never have to look at it again …

ah-d

That would be truly useful. What else have you made?

rp

We've embedded a chunk of electronics in an earring. There is a little infrared LED, which shines through a hole on to your ear lobe. The signal transmitted through your ear picks up the volume of blood flow through your ear, and from that we can get heartbeat information. It's called a photoplethysmyograph. You have a cord running over your ear and into your jacket, where we can hide a computer to pick up the signal. It can take that signal, some respiration information, maybe some muscle tension, some skin conductivity, whatever you feel comfortable providing to your computer, and from that it can try to infer things about your state, including not just things like stress level, but also maybe whether an interaction with the computer is going well, or could be improved.

ah-d

But how is it possibly going to be useful to me, the fact that the computer knows that I'm frustrated or angry or in love?

rp

One of the things that guides a lot of our thinking is looking at how important it is for people to respond to each other's affect. It turns out that a lot of things that are important in human–human interaction are also critical in human–computer interaction. For example, people tend to be nicer to each other face-to-face than behind one another's backs. If you take out a person and put in a computer, people are nicer to the computer face-to-face than behind its back. They treat it as if it has feelings, like a person, even though they know it doesn't have feelings, and it is just a machine. When you ask people later why they did this, they are usually surprised that they did it, and they try to rationalize it in various ways. Dozens of similar studies have shown that the basic rules governing human–human interactions carry over to human–computer interactions.

Think for a second about a particular human–human interaction: suppose you're trying to learn something, and your human mentor gives you some information and you're confused. You sit there and you furrow your brow and you look at it, and your teacher doesn't even notice that you're confused, but just keeps telling you more and more, and you start to get agitated. But instead of recognizing that you are lost and frustrated, the person keeps blabbering on presenting things, until you start to feel like a failure, and want to quit. Right now that's what computers do too. They scroll all this information past you, they present you with the next problem, all they care about is whether you got it right or not; in fact they don't even really care about that. They just keep a record of your errors.

What we're trying to do is give the computer the ability to see, especially in a learning situation, if the student is confused, if you're getting frustrated, if you're interested, if you're bored. Because it's critical to respond differently in those cases. It's not enough to know that the student is making errors. You could be making errors, but looking interested and feeling curious, or you could be making errors and cursing and feeling ready to quit. A good teacher can see the difference. A good teacher knows to encourage now, or to give a more challenging problem, or to give a hint – to mediate the presentation based on your affect.

ah-d

You're assuming now that the computer is acting as a teacher, but I don't use my computer like that; I write. I try to avoid having to learn to use new software; I want to go back and use the simple stuff. But every time you buy a new computer, it gives you a whole lot more stuff, which just makes for more complications.

rp

You're not alone in having that experience. One of the things we're trying to do is measure how aggravating those things are. You can quantify price, you can quantify performance. We have a saying in the US: if you can't measure it, you can't manage it – which is probably true in the UK too. The problem is that right now you can't measure frustration and stress as easily as you can measure price and performance. You can measure how efficient people are in using the software, but you can't yet easily measure the fact that these two software products were equally efficient, but this one drove people nuts. If we could quantify that, put that on the stress scale, I think a lot of people would pay more for the product that was more pleasing.

Too many computer designers have focused on the engineering mentality of just adding more features, adding more bells and whistles, without any regard for the fact that it's increasing your cognitive load as a user, it's increasing your stress, it's actually slowing down your performance, as well as slowing down the machine physically. We all have stories of our older machine that is actually more efficient at doing some job today.

We're trying to help with this. One of the things that we've developed is a sensor than can be used with the ordinary keyboard–mouse interface. It's an ordinary mouse, to which we've added eight pressure pads, which pick up how you're handling the mouse. The mouse already measures what you're clicking on, but it doesn't measure how you're clicking. People handle a physical object differently when they're pleased and relaxed than when they're getting aggravated with the situation. We've been able to measure differences in how people handle the mouse when all is going well, versus when there's some problem, even a small problem.

One of the experiments we did was to have a bunch of our college students come in and put their résumés on a jobs website, filling out web pages; we told them that the whole task should take less than fifteen minutes, and that MIT students all completed it in ten minutes or less, just to put a little pressure on them.

So they use this mouse and put their résumés on the website. On the second page, they fill in where they graduated, the year of graduation, some other information. As they click to go to the third page, it says 'Error, date in wrong format'. That's easy; no matter what date format you enter, we can tell you it was wrong. So then they have to go back and enter the date again; and unbeknownst to them, when they go back to fill it out, the computer has lost their data – something that has happened to everybody. So they have to fill out this page again. We get from the mouse the pressure pattern of people doing exactly the same task twice in a row, one time when all seemed to be going well, and another time when there's a little bit more stress. In 16 out of 17 of our subjects, they applied significantly more pressure to the mouse: there were big peaks in the pressure data the second time through. And this was a pretty insignificant little problem. You can imagine what kind of pressure difference you get when something really important is at stake.

ah-d

Do you get the same pressure difference with all of the pressure pads, or is there some differential?

rp

There are individual differences, and this is why it's important to gather a constellation of information, and to look at the pattern for an individual over time and see where they show their typical behaviour versus when it really goes crazy. However, we found there were some variations that were pretty much person-independent, so you wouldn't need to learn much about an individual; you could just assume everyone was the same.

ah-d

OK, I'm beginning to be convinced that it might be useful for the computer to read my emotions. But now, do you really want to give emotions to the computer?

rp

There are certainly positive and negative aspects to that. A lot of people may want their computer to be able to feel all the pain it has caused them.

But then, I don't want my office computer being surly; I don't want to have to wait for it to feel interested before it's going to listen to me. I don't want it to be an equal to me in that sort of way. At the same time, we've been learning that emotions play a background role that seems to be critical for intelligence, and this presents us with a bit of a conundrum.

Most of us think of emotions as something outwardly problematic. You get emotional, and that's a bad thing; it impairs your judgement, and impairs your thinking and decision making. So we think of the downside of emotion, we think about it like a tornado ripping through life. However, what we don't realize is that emotion is not so much like a tornado – that's just one emotional state – as like the weather in general. Emotion is the background temperature and pressure that is always there. You're always having emotional changes, and if the parts of your brain that control these emotional changes are disconnected through some kind of brain damage, then the person who has experienced this not only appears highly unemotional, but their intelligence is also impaired. In particular their ability to make rational decisions and to engage in social interaction is impaired.

Emotions play a background role that seems to be critical for intelligence.

ah-d
But it didn't happen with Mr Spock and Data.

rp
A lot of people think Mr Spock didn't have emotions, but in fact he did. He was half human, and that's critical, because he had emotions, he was just very good at not displaying them. That's actually a good model for computers. They probably need some internal mechanisms that perform

the important roles that emotions play, but I don't want them displaying them in a violent way. We don't want the mice to bite – although we have talked about having the mouse yelp if you are too harsh to it. There are some indications that for those people who suffer wrist problems, repetitive stress injuries, the mouse can be one of the first points to pick up the fact that they are applying too much tension and so aggravating the problem. We are working with a Harvard School of Public Health expert right now on how to have the mouse give feedback.

ah-d
That'd be good. Now, you've written about HAL, the computer in *2001*. One of the most poignant scenes in the whole movie is when Dave is switching him off, when HAL has finally flipped his lid, and he's saying, 'Please don't do this, Dave', and singing 'Daisy, Daisy', and Dave's unplugging his memory banks; there seemed to be more emotion in the computer than there was in the astronauts.

rp
Yes. One of the points I've written about is that HAL was the most emotional character in the film. We went through the film, looking for examples of human faces showing facial expressions, so that we could run our computer vision algorithms on them. We couldn't find any, except for one surprisingly subtle expression: there's one scene, where Dave is trying to get back into the spaceship, and he's clearly upset that HAL has killed his friend, and that now HAL won't let him back on the ship; so his life is in jeopardy. He's saying, 'Open the pod bay doors, HAL', and you don't see an angry face in the traditional sense, you just see tightening of the lips. Dave is remarkably controlled given the danger he is in. The humans in that film are very unemotionally expressive, even though they certainly feel emotions. You never really see HAL expressing emotions,

although you start to sense what's going on, until at the end he says, 'I'm afraid, Dave', as Dave is pulling out the parts of the computer. You also get the feeling that he's getting cold.

There are also wonderful scenes of them playing chess and of HAL admiring Dave's artwork. It's interesting to ask about the role of affect in aesthetic judgment. Some of us have a hard time trying to articulate for the rest of the world what it is that makes for a successful piece of art. To what extent is it a cognitive experience, or a perceptual experience, or an emotional experience? The fact that HAL could appreciate not only the fact that the drawing was a realistic resemblance, but also that it had some other beautiful aesthetic qualities, shows that HAL might have had some of that emotional appreciation.

ah-d
Do you think that was Arthur C Clarke, or was that Stanley Kubrick?

rp
My guess is that was more Kubrick. If you look at the book versus the film – where Clarke was first author of the book – the emotions were downplayed. It's all cold, cognitive judgment – HAL sensed that there was this discrepancy – that led to the problems in the film. In the movie, you don't hear what's going on inside HAL's head; you just see the emotional behaviour and the emotional response at the end. The book shows more of HAL's thought processes, so he seems more rational.

ah-d
The book was actually written after the film, wasn't it? They wrote the screenplay together and then Arthur pulled it out to make a book. It's rare that you can lay book and film side by side and look at them.

rp

Yes, and see the two different personalities contributing, like a tug of war between the emotions and the intellect. This was an interesting tug, because if you disable the emotions and let the intellect reign, you might think what you'd get would be highly rational and highly intelligent, but it would be a major flop, and not just as a film, but in real life.

For example, Data, the android character in *Star Trek*, is very misleading to people. The science there is much worse than it was with Spock. Data is a creature that can click off his emotion chip and still function in a highly rational way. But from what we know about the human brain, when you click off the connections between the frontal lobe and the amygdala and these lower limbic structures, you actually cease to function in a highly rational way.

ah-d

Can you explain that? I find it hard to believe that you need emotions to make rational decisions. Often emotion gets in the way of rational judgement.

rp

Again, it's surprising because it's so subtle; it's in the background. Most great scientists would say that the way that they got to the key insight in a crucial proof was not via some logical chain, but via an intuitional path, and that they only later went back and filled in the logical chain. You can think of these intuitions as background processes that might actually be quite logical, happening in the brain. But usually we're not really aware of them. It's only later that we realize, yes, I was feeling that that was the right thing to do, and now I can articulate why. These background processes seem to be important for dealing with a lot of complex, unpredictable inputs in a finite amount of time. We just don't have time, logically, to address all the possibilities in the time that it takes to say, yes, I'll see you at five o'clock today.

ah-d

But it still doesn't seem to me that it's emotion leading you to make that jump.

rp

I think that we think about emotion in a very limited way. Emotion, I think, would be more appropriately thought of as temperature and pressure and humidity, which together make weather. Every now and then these components arrange themselves in such a way that you have a gorgeous sunny day, a state of great joy, or an oppressive sadness, pouring rain. But those are the extremes. Typically, you're in some kind of a neutral state, and so what is the role of all those background signals? Well, they're regulatory, and they're biasing, much like the heating, ventilation, and air-conditioning system in a building. You don't usually think about it, unless it gets out of control and your office is too hot. Ordinarily, if it's doing its job, you don't even notice it; it's in the background. But it's critical for keeping things comfortable, on track, regulated.

We aim to not make emotional machines that would go out of control, like HAL did, but machines that have the skills of 'emotional intelligence'.

ah-d

All right. You're beginning to convince me. You're not there yet. Suppose you were able, in the media lab, to make HAL: would you want to do it?

rp

Not in quite the same way. We would love to be able to have a computer see as well as HAL could see, and talk and understand language as well as HAL could. We already have computers that can play chess. What we would want, though, is a computer that, if it had emotions in the way that HAL is presented as having them, also demonstrated some emotional competence with respect to handling crises. For example, there's no evidence that HAL had ever been put through the mill in handling a crisis.

Let's go back to the human–human situation. If you're going to put a person in an extremely important role, heading a company or a government, you want to make sure that that person has been through a lot of difficult times and proven that they can handle these crises while remaining strong and capable. HAL was presented with a dilemma, and just went berserk, started killing people, all in the name of protecting this mission. Intellectually, HAL was a giant; but emotionally, he was a juvenile. In a sense he portrayed our limited understanding of emotion at the time: emotion as something that makes you irrational.

Decades later we know that in an intelligent healthy human, emotion is something that contributes significantly to your rationality. Our challenge today is to create machines that have a balance – mechanisms of emotion that are important for intelligent functioning, and the proven ability to regulate these emotions in service of higher goals. Thus we aim to not make emotional machines that would go out of control, like HAL did, but machines that have the skills of 'emotional intelligence' – that can recognize and respond intelligently to emotion, and utilize such mechanisms for fostering more natural and intelligent interaction. Someday when you interact with a machine like this it shouldn't seem emotional at all – if implemented right, the emotional capabilities will not draw your attention. Instead, the machine should just seem more useful, helpful and enjoyable. We will have succeeded when ‘affective’ becomes nicely confused with ‘effective’.

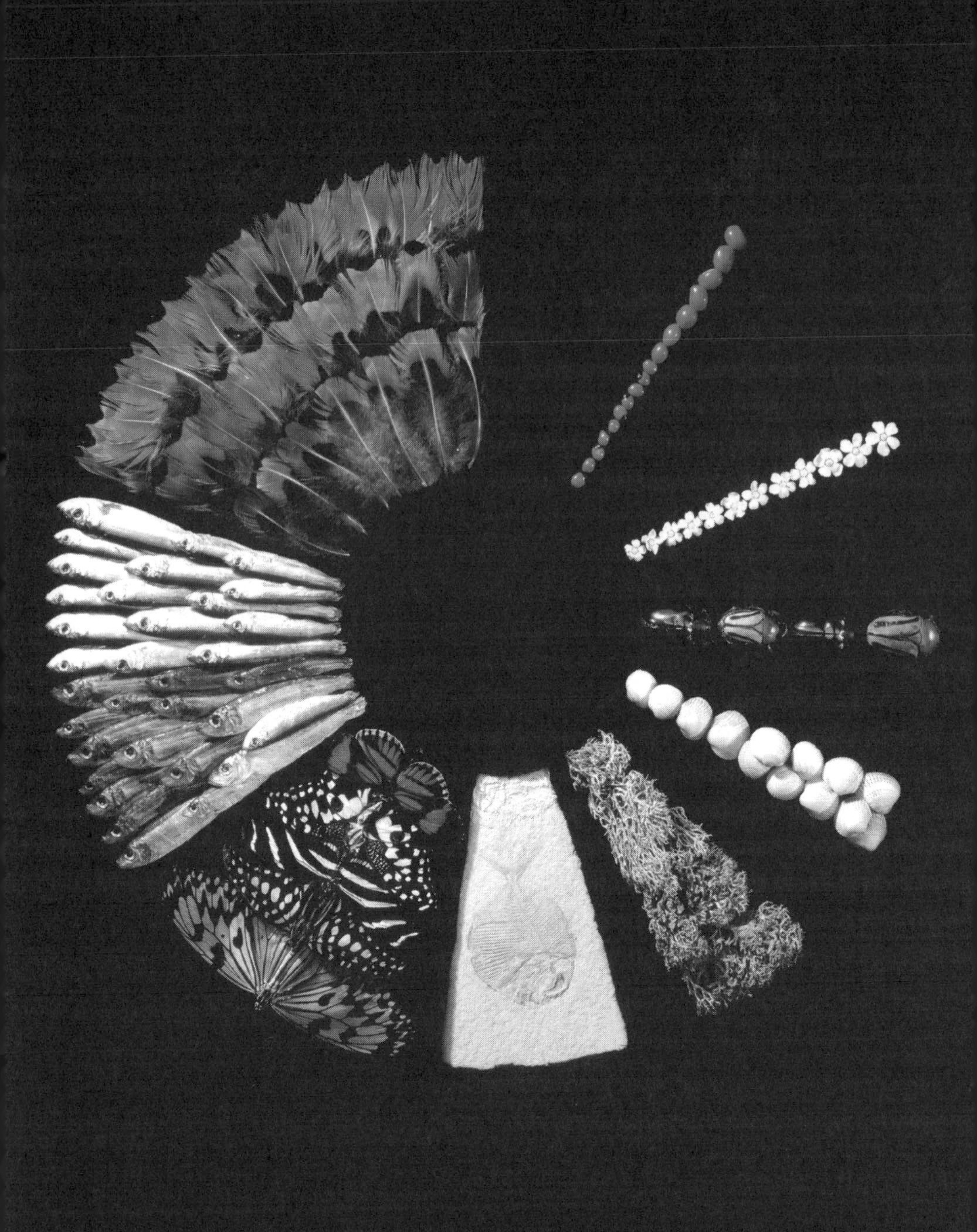

chapter ELEVEN

Peter Raven

Biodiversity

Peter Raven was born on 13 June 1936 in Shanghai, China. He received his PhD from the University of California at Los Angeles, after completing a degree at the University of California at Berkeley. In 1971 he moved to St. Louis to head the Missouri Botanical Garden, turning it into a leading tropical plant research facility, cataloguing species close to extinction, and providing not only horticultural displays but also educational programmes. He was a member of the President's Committee of Advisors on Science and Technology during the Clinton Administration, and in 2001 received the National Medal of Science. He is now Professor of Botany at Washington University in St Louis and, as a leading advocate of conservation and a sustainable environment, champions research around the world to preserve endangered plants. He is co-editor of *The Flora of China*, a joint Chinese–American project to create a contemporary account of all the plants of China.

ah-d

Time magazine has called Dr Peter Raven one of the heroes of the planet. Peter, one of your great concerns is biodiversity. Tell me, what is biodiversity?

pr

Biodiversity is a contraction of biological diversity, but what it's come to mean are all of the living things on earth: plants, animals, fungi, microorganisms, all of their genetic variation, and the communities and ecosystems that they make up. In other words, all life on earth – the web of life – that's what biodiversity is. It's being seriously threatened by activities of human beings, largely over the last few hundred years. Since the great extinction 65 million years ago at the end of the Cretaceous Period, when the dinosaurs disappeared and mammals came into the ascendancy and so forth, life in general on earth has been increasing in magnitude and diversity. If you look at the fossil record to see how long individual species have lasted on average, you find about one species per million per year becoming extinct.

If you look at the rate for the last few hundred years, in other words largely during the Industrial Revolution, when we've known enough about some groups, such as birds, mammals, butterflies and plants, you find that more like a hundred or a thousand species per million per year have become extinct. The destruction of habitats all over the world is going on so rapidly that this number is likely to become very, very much higher. In other words, we might see the disappearance of a huge proportion of all life on earth during the course of the century that we've just entered.

We're losing thousands of species per year, right now, as the habitats get cut back. Tropical moist forest, for example, which contains a very high proportion of all the species on earth, may be down to about 5 per cent of its original extent by the middle of the century. If that happens, and we don't pay much attention to the process, we could lose two out of every three species by the time 2100 rolls around. That would be a rate of extinction comparable to the one that happened 65 million years ago. But there is in fact lots that we can do about it.

The first problem is to calculate roughly how many species there are on earth. Lord May has tried to estimate this. Start by saying there are about 1.6 million that are actually described, and have names. Bob's estimated that there are between seven and thirteen million actually existing. Many people think there are a lot more. But it turns out that there are only a few groups, for example mammals or birds or plants, where we have a pretty good idea of the total number of species. For others, we've barely begun to scratch the surface. We really don't know how many there are; we can only make estimates by saying, well, go take another sample, how many new ones were there? How many did you find that you'd never seen before? That's how you get to estimates like ten million. But other people would say there might be tens of millions. The practical point is that, while they're becoming extinct, the majority of what is disappearing will consist of species that we've never seen, and whose existence we've never known about.

ah-d

We seem to be fantastically ignorant. I'm sort of shocked by that. Why is that? There have been naturalists around for hundreds of years.

pr

Victorian era naturalists wandering around with butterfly nets just didn't do the job on things like fungi worldwide, or nematodes, small worm-like animals that live in the soil in vast numbers. Many other groups have been badly neglected: there are hundreds of thousands of kinds of mites in the world, people would estimate, and yet only 35,000 have been named.

ah-d

So you say we're killing perhaps two-thirds of the species on earth? Why does it matter? Surely we need only a hundred species or so, cows and sheep and so on. Why do we need all these other species?

pr

There are several reasons. You can look at species individually and point out that all of our food comes from plants, directly or indirectly. Most of our food comes from just over a hundred kinds of plant, and in fact 60 per cent of our food comes from three members of the grass family, maize or corn, wheat, and rice. But, on the other hand, there are tens of thousands of different kinds of plant that have been used as food at different times in human history – out of the estimated total of 300,000 – and if we really want to be able to grow them better and in a more adept way throughout the world, we'll want to use more than the 100 that we depend on now.

Then medicines. Most of the people in the world use plants directly as sources of their medicines. Even for those of us who depend on prescription drugs, lots of those prescriptions are for natural products that are now manufactured – like aspirin, which is just a slightly modified form of a natural product – or that have been improved by working with natural products along the way.

If you consider what we've come to understand about molecular biology over the last 50 years, and the ways in which we can shift the characteristics of organisms by gene-splicing and by other techniques that are just coming on stream, and then you consider what power genomics – knowing about the whole genomes of organisms – has given us, you realize that we're just beginning to get the tools to use individual kinds of organisms in much more complex ways than we can imagine. It's not a good strategy to be losing so many of them just as we're learning how to understand and utilize their characteristics. Putting it another way, it's been said that the twenty-first century is likely to be the age of biology, just as the twentieth century was the age of chemistry and physics. But if we're going to make an age of biology out of it, we have to have some biology to do it with.

In other words, when you add it all up, losing biodiversity is not very intelligent. And of course, it's irreversible; once a species is gone, it's gone forever. We not only deny our descendants the right or the ability to use those organisms and knowledge about them, we also lose a lot of the properties of organisms by which they combine into ecosystems. These provide what are called ecosystem services, free services, such as protecting soil, protecting watersheds, making rich breeding grounds for marine organisms along the edge of the sea, providing pollinating insects for our crops, and all of those community functions that we also understand pretty poorly. We're having to restore a lot of the world; we've hammered a lot of it pretty badly. We'll have to restore a lot of it in the future, and in order to do that we need a diversity of organisms.

Finally, you might say there's just sheer enjoyment. A world with fewer organisms is simply a lot less interesting, a lot less pleasant, a lot less beautiful, peaceful and healthy than a world in which there are many.

ah-d

You've painted us a fairly gloomy picture of vanishing biodiversity, but you've said there are things we can do about it. What can we do?

pr

A high proportion of the world's biological diversity is concentrated into a relatively small percentage of the world's surface. Areas like the Cape region of South Africa have enormous concentrations of plants; the whole area around the Northern Andes of South America – all the forests on their slopes and around them – is estimated to have about one-sixth of the world's biodiversity. If you really were to preserve those areas, to set aside the undisturbed part of those areas, you would be able to save a lot more biodiversity than if you did nothing. Organizations such as Conservation International are now trying to focus on these, and to put together the money that it would take to protect them.

A high proportion of the world's biological diversity is concentrated into a relatively small percentage of the world's surface.

Returning to the Northern Andes, the three countries of Colombia, Ecuador and Peru have an area about 30 per cent of that of the continental United States, but, as I mentioned, about one-sixth of the world's biodiversity – an amazing concentration of life. It's poorly known, and the landscape there is being altered rapidly.

ah-d

Is this a consequence of the geochemistry? Of the various sorts of soil and so on?

pr

If you consider Africa and Asia, it's the only major mountain range running north–south through the tropics. The Andes have been uplifted primarily in the last few million years, with lots of opportunities for the evolution of new species at different elevations while they were being uplifted, so there is a great concentration of habitats in a very small area. When you think about temperate mountains you think of animals migrating up and down them at different seasons, and adjusting to them and so on. But tropical mountains are not like that. They're constant in temperature at different elevations, so the net effect is that you have a lot of different habitats, quite distinct from one another, packed close together. The soil is richer around the mountains too: the fertile, relatively recent soil of uplifted mountains makes for a lot more opportunity for biodiversity than somewhere right out in the middle of the Amazon Basin, which has, by and large, much less fertile soils.

ah-d

You said we can protect these hot spots of biodiversity. What else can we do? Can we use technology?

pr

Different things can be done with different groups of organisms. For example, there are 300,000 species of plants, maybe ten per cent of which we haven't found yet, but many of which we know. This sounds impossible to deal with, but it might surprise you to know that an estimated 85,000 of them are already in cultivation in botanical gardens. Sometimes that's just one individual, sometimes more, so it may not be genetically very viable but, on the other hand, cultivation out of the field for threatened and endangered plants is one good option for their conservation. An organization called Botanic Gardens Conservation International, with its headquarters at Kew, near London, is an international network of botanical gardens to which we in Missouri belong, and which we support and interact with at many levels. They are pursuing, among other things, a worldwide scheme through which we can save a major proportion of all of the plants in the world simply by collecting them from the wild and bringing them into cultivation. This has now received international backing through the United Nations Convention of Biological Diversity Secretariat.

Plants can be kept in cultivation, or as seeds in seed banks where, at very low temperatures, around –100°C, the seeds may last for a hundred years before they're brought out again. It's obviously a lot easier to keep plants in those circumstances, or in botanical gardens, than it is to keep a vertebrate animal, where you have to maintain a viable breeding population. So in a relatively cost-effective way, it's possible to preserve many of the plants of the world away from the natural areas where they may be in danger of extinction.

There are more than a quarter of all plant species already in cultivation now, and it is relatively easy, if the time and money are available, to bring many more into cultivation. We in Saint Louis have a big collection in certain groups – for example, in the aroids there are the philodendrons, anthuriums, calla lilies and their relatives. We have thousands of strains, individual species, largely from Latin America; many are from places where the original forest habitat has been cut down, and they might or might not still exist in nature. It's hard to tell when you're dealing with tropical moist forest, because some little patches might or might not have the individual species that were in it in the first place.

ah-d

So the fact that you provide a beautiful environment for people to wander around in, as they do at Kew, is only incidental to the scientific work going on?

pr

Yes and no. This is how people get educated about plants, by looking at them and experiencing them. We gather plants, we try to conserve them, we do research on them, we spread the results of that research, and we try to show the public how beautiful and interesting and diverse they are by the displays that we have on the grounds and in the greenhouses. It's about the same as in a natural history museum, you might say, where the exhibits by and large will present a lot more of the facets of life than what is actually going on behind the scenes. But it's the relationship between the research and the exhibits and the education that's important, and that makes these institutions such key institutions in learning about, disseminating, educating people about, and demonstrating the beauty and the diversity of plants.

ah-d

Now, I am told that you built this place up. You took it over when only a small part of the area was developed as a botanical garden, and many of the buildings and other facilities were badly in need of repair. You've trebled the cultivated area; you employ a mass of people; and you've added greatly to the display collections and to the scientific staff. Are you a gardener?

pr

No, I'm a plant scientist, plant classification and evolution.

ah-d

So you love the plants theoretically, rather than wanting to get your hands dirty?

pr

I like to get my hands on them, so that I can press them and make them into herbarium specimens for study. I love living plants and am deeply interested in learning more about them scientifically and in applying that knowledge for human benefit.

ah-d
And you travel the world. You've written a book about the plants of China.

pr
I'm the co-editor of a major project revising the account of all the plants of China. About a tenth of all the kinds of plant in the world exist in China. If you look at the fossil record of the whole the Northern Hemisphere in the middle of the Miocene Period about 15 million years ago – the plants in Europe, North America and China were roughly the same. But if you look at them now, you see Europe, China, and the United States are each about the same size, but Europe collectively has fewer than 12,000 species of plants, the United States about 20,000, and China about 30,000. In other words, they've survived in China; so it's a great laboratory of plants that have been more widespread in the Northern Hemisphere, but which have survived only there.

The reason, presumably, is that in China the plants have migrated north and south as the climate has shifted and become more seasonal from the Middle Miocene onward – the last 15 million years – whereas in Europe they come smack against the Mediterranean and ultimately the deserts. In North America, similarly, they are cut off by the Gulf of Mexico. In China, though, the ability to migrate north and south seems to have given a lot more opportunities for survival. And that's why China's plants are so interesting scientifically, and provide such a good link with the past. And of course, if you think about horticulture primulas, rhododendrons, forsythia and so many of our garden plants, as well as so many of our economic plants – rice, soybeans and so forth – come from China in the first place. It's a particularly fascinating area to study, and a fascinating experiment in human relations, too, to cooperate on that scale.

The project's just about a third finished, and we're hard at work on it now. We have many cooperating institutions and individuals.

ah-d

I remember when I'd just finished my PhD, the Club of Rome commissioned some people to look at the future of resources on earth. The result was a series of dire predictions: that manganese was going to run out in 1986, copper was going to run out in 1990, oil was going to run out, and by 1995 the whole world was going to be dead. They called the book *The Limits to Growth*, but they were wrong.

pr

The most important thing about that study is that it was the first effort to provide a model for what was happening to the whole of earth. Before that, nobody had been able to think about it as broadly as they did. It was very valuable in that way, and it was written up in early reviews as an outstanding study and very important for that reason, as it should have been. On the rebound, nobody in 1972 really wanted to hear that there were limits to growth. It was a bad phrase. The reaction was, no, oh no, that can't be happening.

The other thing that's happened, which was really quite unpredictable, is that in those days, people tended to think that minerals, oil and things like that were running out very rapidly, whereas productive ecological systems, agriculture and so on, would go on for ever. In fact what's happened is more or less the opposite. The supplies of the things that are really obviously fixed, like magnesium, iron, copper and petroleum, seem to be larger than predicted, but the renewable sources of energy, like natural forests and agricultural fields and so on, are being hammered so hard by human beings that it's they that are collapsing.

But going back to the original study, the important thing was that it was the first model. Being a model with explicitly stated terms, those terms could then be changed as more knowledge was accumulated. There was a misunderstanding of the nature of the use of non-renewable resources, but the important thing was the attitude.

ah-d

Now you're saying some similar doom-laden things about the future of the planet. Could you be wrong?

pr

Anybody can be wrong about the future. You can use various kinds of scenarios and predictions and models, but the important thing is, as in all of science, to be able to make hypotheses and modify them as new knowledge becomes available. It is, however, a matter of record that over the past 50 years, 20 per cent of the world's topsoil has been lost, and 20 per cent of the world's agricultural lands are no longer available because of urban sprawl, over-fertilization, desertification and other factors. One-sixth has been added to the carbon dioxide in the atmosphere, which is the major greenhouse gas that's driving global warming. Seven or eight per cent of the stratospheric ozone layer has been depleted. A third of the forests that were around in 1950 have been cut down without being replaced.

Those are all demonstrable trends, as is the trend in species extinction that I talked about earlier. In 1950, the global population was about two and a half billion people. Now, 50 years later, it's about six billion people. The extra three and a half billion people are obviously using up a lot of these resources and causing environmental changes at an increasingly rapid rate. If one projects these trends forward – and the median World Bank estimate is for about another three billion people before world population levels off – and one thinks about growing levels of affluence and therefore of consumption all over the world, it's obvious that more and more damage is likely to be done. In other words, we do need a levelling-off of the population, which requires continuing human effort. We do need a rationalization of the levels of consumption of the world's natural resources. And we do need new technologies to help us out of the dilemma that we've entered.

ah-d

There are real problems with energy, aren't there? What proportion of the world's energy is used by the United States?

pr

We produce about 30 per cent of the carbon dioxide, is one way of looking at it. We use about 25 to 30 per cent of the industrial energy. One of the reasons the industrialized nations of the world, which have 20 per cent of the world's population, are called industrialized is that they use such a high proportion – 80 per cent – of the world's petroleum, natural gas and coal. That means that they emit a corresponding proportion of the world's carbon dioxide.

A world with more than six billion people in it is a world where we are going to have to act collectively, whether we want to or not.

ah-d

But we in the industrialized nations don't have the right to prevent the developing nations from industrializing, do we?

pr

Of course not but, on the other hand, we have a collective interest in how we develop together and what the effects of particular kinds of development will be on the sustainability of the earth. A world with more than six billion people in it is a world where we are going to have to act collectively, whether we want to or not. In other words, the kinds of dreams and aspirations that people had when they were founding the United Nations are going to have to become realities, because six billion people, seven billion people, nine billion people can't simply live together as if there were no limits to growth. There *are* limits to growth. And in order to deal with them, we have to deal with them collectively, at some point. We have to get to grips with it.

No, we have no right to prevent growth around the world but, on the other hand, there's not enough room in the world, if we use present technologies, for everyone to live at the level of the United States. In fact, it's estimated that if we all did, right now, it would require two more planet earths to house the people, and I don't know where we're going to get them.

ah-d
Are there simple things we can do? Should I recycle, for instance? What are the simple things that individuals can do?

pr
There are several different kinds of things, and they can be expressed in different ways. I think one of the most important things for people who live in industrialized countries to do is to encourage an international spirit among the citizens. It sounds very simple-minded, but the fact of the matter is that 80 per cent of the people in the world live in developing countries, and until we get to know, understand and appreciate them, and accelerate the importance of that in the minds of our fellow citizens and thus in the minds of our governments, we can't possibly reach a kind of a collective stability because nobody's going to care.

It's notorious in the United States that when you're talking about Third World news, major disasters involving thousands of people are necessary to attract any kind of attention and, correspondingly, that much smaller events in Europe or the United States grab the headlines. We've got to understand people around the world, and anything that we can do as individuals to encourage that will be very important. But yes, limiting our own consumption and particularly not being wasteful in our own activities is crucial. The United States uses about twice as much energy per person to support our society as people do, for example, in Germany, Switzerland or Sweden, and yet the standard of living is about the same. The only way you can figure that out is there's a lot more waste in the United States than there is elsewhere. The amount of garbage produced shows the same thing. So, yes, I've taken a long path getting to the answer to your question but definitely do recycle to the extent that you can, and encourage others to do it too.

ah-d
You believe in big corporations. You think they are better people to deal with than governments because they're not just waiting till the next election.

pr
I would say they are necessary to deal with. They're an economic force that exceeds the economic force of governments and multinational institutions taken together. Obviously governments and multinational institutions have to provide a framework for all of this, and individual nations take care of their people through a framework of laws, and in a way we do this indirectly through multinational institutions. But the corporations that understand and take on board global environmental trends are likely to be the corporations that are going to be profitable and return value

to their shareholders in the long run. It's not a matter of cleaning up their act and doing things for altruistic reasons; it's a matter of understanding the realities of the world as they are, and acting on them in such a way as to be self-serving. Because of the ability of corporations to do that, which is being demonstrated in an increasing number of cases, I think that they will, as they must, become a major global factor in all of this.

ah-d
This situation is obviously quite worrying. Are you an optimist or a pessimist?

pr
I always say that it's wrong to be an optimist because of wishful thinking. You can say, well, maybe a rocket ship is going to land from Mars and give us an infinite source of energy. Maybe science or engineering is going to come along with new inventions that will make everything all right. But the scientists and engineers who've considered these possibilities don't think they're going to happen, so it's perverse for economists to say science and engineering is going to save us, when scientists and engineers don't believe they can.

If on the other hand you believe that you can do something about the situation individually, then I think you have a reason to be an optimist because of what you intend to do. The world is clearly running downhill in some sense, environmentally becoming less diverse, more homogeneous and less interesting than it has been. But, having said that, it has to achieve a stable level, a sustainable level at some point, simply because there's only so much productivity there.

What I see us all doing collectively, whether we recognize it or not, is determining where we'll reach that kind of equilibrium. At the end of the day, the world's likely or certain to be a patchwork, some places more beautiful and interesting and nourishing and healthy than others. How those places turn out very locally, well, we're going to determine that individually by what we do; so let's try to do a better job – and for a start let's do our best to preserve as much biodiversity as possible.

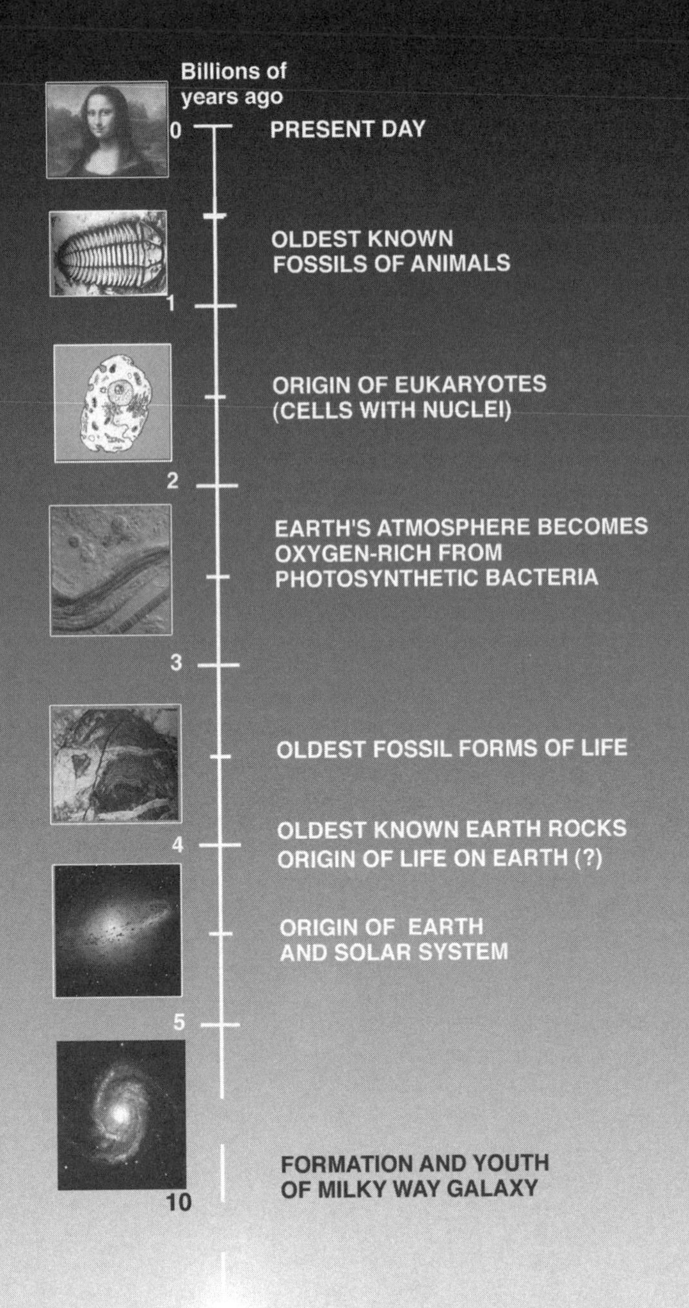

Billions of years ago
0
PRESENT DAY
OLDEST KNOWN FOSSILS OF ANIMALS
1
ORIGIN OF EUKARYOTES (CELLS WITH NUCLEI)
2
EARTH'S ATMOSPHERE BECOMES OXYGEN-RICH FROM PHOTOSYNTHETIC BACTERIA
3
OLDEST FOSSIL FORMS OF LIFE
4
OLDEST KNOWN EARTH ROCKS
ORIGIN OF LIFE ON EARTH (?)
ORIGIN OF EARTH AND SOLAR SYSTEM
5
FORMATION AND YOUTH OF MILKY WAY GALAXY
10
14 (APPROX.)
BIG BANG BEGINS THE UNIVERSE

chapter TWELVE

Sir Martin Rees

Where we are in the universe – or in one of the multiverses

Martin Rees was born in England on 23 June 1942. After studying mathematics at Cambridge University, he held post-doctoral positions in the UK and the US before becoming a professor at Sussex University. In 1973 he became a fellow of King's College and Plumian Professor of Astronomy and Experimental Philosophy at Cambridge. He is now Professor of Astronomy and Cosmology and Master of Trinity College, Cambridge, and holds the honorary title of Astronomer Royal. His current research deals with general cosmological issues, and more specifically high-energy astrophysics and cosmic structure formation. His latest book is *Our Final Century.*

ah-d

Martin, as Astronomer Royal, do you actually go along to the Queen with a pair of binoculars and say, 'Have a look at Mars?'

mr

I don't. It's really an anachronistic title that dates back to the seventeenth century when the earliest Astronomers Royal were concerned with longitude and the calendar and navigation. Now it's a purely honorary post, but it does reflect the fact that astronomy was the first professional science apart from medicine, almost certainly the first to do more good than harm and, back in the seventeenth century, it was supported by the public and by the government.

ah-d

Ah, you don't get invited to tea at Buckingham Palace?

mr

No, I don't, and the pay is zero. I do have a day job, fortunately, as a Professor of Astronomy at Cambridge University.

ah-d

Good. Now, in your book *Just Six Numbers* you lay out a recipe for creating a universe, if you like. Could you have written that book ten years ago?

mr

I don't think I could, because we are living in a remarkable time in cosmology, a time when the big picture is coming into focus. Perhaps I can give an analogy with understanding the earth. Five hundred years ago we didn't really know the layout of the continents, the size of the earth, or its exact shape. Through the discoveries of great explorers, we learned the general configuration of the continents, and now we don't expect any big surprises. In the last ten years, we've really got to that stage for our entire observable universe. We know roughly how big it is, roughly what shape it is and roughly what its basic ingredients are: galaxies, different kinds of matter, different kinds of energy and so on. So this is a special time in our understanding of our cosmic environment.

ah-d

But you cosmologists just sit around in armchairs thinking. How have you got this sudden understanding so recently?

mr

Well, certainly not through pure thought. In fact, by sitting and thinking, we wouldn't have got very far, because we are no wiser than Aristotle was more than 2000 years ago. The reason we've made progress is primarily through technical advances, through the fact that we can, with telescopes and other kinds of probes and other kinds of experiments, learn about the quantitative features of our universe. With the most powerful modern telescopes, on the ground and in space, we can study objects so far away that we're seeing them as they were a long time ago, because light takes a long time to get to us from immense distances. We can see objects as they were ten billion years ago. We have a sort of time machine.

We have evidence as to what the universe was like in the first second of its history. We believe, with great confidence, that everything started off with a so-called hot big bang, and we can trace out the subsequent evolution, at least in outline. And this broad cosmic perspective has come into focus within just the last few years.

ah-d

How confident are you that you're right?

mr

Well, not completely confident, because I'm always mindful of two things. First, Ptolemy thought that by adding more and more epicycles, he could fit any features in the solar system. And also there's a well-known dictum that cosmologists are often in error but never in doubt. For those reasons I think we should keep an open mind. But, nonetheless, having given you that sort of health warning, I would say, with 99 per cent confidence, that we can extrapolate back to when the universe was one second old, and at that stage everything was squeezed hotter than the centre of the sun. The temperature was about ten billion degrees – a very hot soup of hydrogen and helium gas, and radiation, and other particles.

There's a well-known dictum that cosmologists are often in error but never in doubt.

But that progress brings into focus a new set of questions: what happened in the first millisecond, even in the first tiny fraction of a microsecond? And there we can't be nearly so confident. Back at one second, although conditions were vastly different from our present universe, they weren't all that extreme compared to what we can simulate in laboratories. The density was no more than that of air, and the particles moving around were at the sort of energies we can easily produce in the lab. But as we go back in imagination to the first tiny fraction of a second, we lose our foothold in experiment. We get to conditions so extreme that we don't know the relevant physics, and that's why, although we've made this immense progress, the frontier with the unknown still exists. We can push back beyond the time when the solar system formed, before the time when the galaxy formed, right back to the first second of the expansion of the Big Bang, where everything started. But in that first tiny fraction of a second, there still lies a mystery.

ah-d

You've said so much there that I want to ask ten questions. But, anyway, you said there were hydrogen and helium after one second. Nothing else?

mr

Nothing else, apart from exotic particles, which now make the so-called dark matter in the universe, and a lot of radiation. But one of the most remarkable things, and this was indeed the theme of my book *Just Six Numbers,* was that from this simple recipe an early universe that can be described by just a few basic numbers can evolve, over 13 billion years or so, into an amazingly complex cosmos. We now look around us and we see galaxies, each made of stars; around many of those stars there are now planets known to orbit; and on at least one of those planets around at least one star. We know there's been this amazing biological evolution leading to creatures able to contemplate their origins, like us. And the question is: how, from these simple beginnings, did the universe evolve its tremendous complexity? Was there something special about the way the universe started to allow this to happen?

ah-d

And you say that these six numbers were all critical; that if they varied slightly, we wouldn't be here?

mr

That's right. I think there are two things we realize from studying cosmology. One is the links between the very large and the very small, the cosmos and the microworld. To understand ourselves in the everyday world, we know we have to understand atoms, how they stick together to make molecules. That's chemistry. But we also need to understand the stars. The reason for that is that everything started off with hydrogen and helium. The atoms of carbon, oxygen, silicon and iron that are crucial to our existence, they were made in stars – forged inside stars as a by-product of the nuclear power that kept those stars shining.

Stars are fusion reactors kept going by nuclear energy, and when the stars die, they explode, or blow out their material, and then new generations of stars condense from that debris. So, our sun was not one of the first stars in the universe. It formed, along with the planets of our solar system, from gas already contaminated by the debris from earlier generations of stars. So, we are quite literally the ashes of long-dead stars, or, to be less romantic, we are the nuclear waste from the fuel that kept these old stars shining. And so, to understand such a simple question as 'Why is carbon so common on the earth but gold so rare?' we have to think about the fate of stars that died far away and five billion years ago, before our solar system formed. That's the link between the cosmos and the microworld, and that's the first thing that strikes us.

But the second thing that strikes us is that for this immense complexity to have evolved, it looks as though the basic laws governing the universe must have been in some respects rather special, because you can easily imagine a universe where the recipe is defined by slightly different choices of the numbers, and this imaginary, different universe would be sterile, or still-born. It might, for instance, re-collapse after a year, before there was time for anything to happen. It might contain just dark matter, particles and no atoms, no chemistry. It might have no force of gravity. It might have a force of gravity so strong that it would crush anything as big as us. Or it might have atomic nuclei in it that didn't stick together to make elements like carbon and oxygen. So we can imagine how the universe could be different. And, indeed, it looks rather surprising that a random universe should have this particular choice of numbers, which allows us to evolve. And the question is: how do we react to that surprising circumstance?

ah-d
You're almost suggesting that it was built for us?

mr
Well, that's one reaction. I think there are three reactions one could give. One is to say it's just happenstance or coincidence; it's a brute fact; we shouldn't worry because we wouldn't be here otherwise. I don't think that's enough; I think we need some deeper explanation than that. There's a rather nice analogy given by the Canadian philosopher John Leslie. He says, 'Suppose you were facing a firing squad. Ten marksmen aim at you. They all fire. They all miss.' Now, it's true that if they hadn't all missed, you wouldn't be there to worry about it. But still, you'd ask why did you survive. You'd think there must be some deeper explanation. Likewise, I think we do need some explanation for this apparent fine-tuning.

Well, some people, of course, say it is Providence. They use the argument familiar from natural theology, Paley's classic argument: just as a watch implies a watchmaker, a design implies a designer. That's an argument that was popular 200 years ago, but in the biological world that argument fell from favour with the advent of Darwinism. And I don't really want to invoke it now, although some people do.

I want to use a somewhat different analogy, actually, which is not the watch and the watchmaker but the ready-made coat store with a large stock. If you imagine going into a shop with large stock of suits, you'll find one that fits you. Likewise, if there were not just one Big Bang, but many big bangs, then it wouldn't be surprising that some of those big bangs were described by the right recipe. And so, if we have a still grander concept of our cosmos and say that the cosmos is a sort of multiverse, not just what we normally call the universe but countless different universes, each starting with different big bangs, and we imagine these all popping off, each described by a different recipe, then most of them might, indeed, be still-born and sterile. But there'd be some where the numbers were tuned in such a way that complexity could arise. And of course, we're in one of those. So, if we invoke this still grander cosmos of many universes then it becomes less surprising that we should be in one where there are fine-tuned numbers.

ah-d

Are you suggesting that these multiverses are sequential, or at the same time?

mr

Well, I'm suggesting they really exist, and there are theories that do suggest that our Big Bang may not be the only one. It might be part of some grander ensemble. As to the temporal sequence, I have to give a rather shifty answer. That is because, as we extrapolate back towards the beginning of our Big Bang, where, as I said, we don't understand the nature of the laws, most people suspect that we have to jettison common-sense notions like the idea of three-dimensional space and one-dimensional time just ticking away. We may have to invoke extra spatial dimensions, and the whole idea of space and time may be more complicated. Indeed, right back at the beginning, we have to worry about quantum effects, quantum fluctuations, making the entire universe fuzzy. So, if we can't think about time in terms of a clock ticking away, we can't really talk about before and after. So we can't really say whether these universes are before or after or, as it were, alongside our universe.

There's one idea, for instance, that there is a fourth spatial dimension and that there could be other universes alongside ours, just like in the analogy of an unbounded sheet of paper with bugs crawling over it, where the bugs think they're in an unending two-dimensional universe. You can imagine two different sheets in our three-dimensional space, and the bugs on one sheet wouldn't be able to envisage the other one. Likewise, there could be other universes separated from ours in a fourth spatial dimension, and we could not envisage them – even if they were only a millimetre away – if that millimetre were measured in a fourth spatial dimension and we are imprisoned in just three. There may be more to space-time than what we call our universe. But it doesn't make much sense to say whether it's before, after, or alongside. There are lots of different theories, and this is all very speculative.

We are quite literally the ashes of long-dead stars.

I was confidently going back to one second and emphasizing that we did, I think, have reasonable ground for taking that part of cosmology sincerely – indeed, as seriously as you should take what a geologist tells you about the early history of our earth, because we have good fossils at that stage. We should not take anything cosmologists say about other universes or about the first tiny fraction of a second of our universe without a large dose of scepticism, because we don't really have any firm guidance. Lots of speculation, though.

ah-d

You're talking about multiverses as though they exist. Isn't this just a convenient cop-out? Because otherwise there is no good explanation for why we're here.

mr

Well, it's a cop-out in a sense. But I would say it is really part of science rather than metaphysics. It's speculative science but it is in principle testable, in two ways. First, we could in principle see if our universe has the properties that it would have if it were part of an ensemble, because of the number of ways in which it would be special but not too special. I discuss that in my book.

But the other way to put this concept on a firmer foothold would be to develop a theory that allowed us to describe what material was like at the incredible densities in the first tiny fraction of a second. If theory explained things we could observe but which are still deeply mysterious, like why there are electrons and protons and why they have the masses they do, and why there are the particular forces of nature we have, then such a theory would acquire credibility. And then we would take seriously any predictions this makes, even in domains we can't observe.

To give an analogy, we've got confidence in Einstein's theory of gravity, because it accounts for the properties of space, time and gravity everywhere we can check it. We are therefore disposed to take quite seriously what it says about the insides of black holes, which we can never observe. Likewise, physicists talk about quarks, which are particles inside an atomic nucleus we can't observe, but whose postulated existence helps to explain things we can observe. And so, I think, we may be able to put the multiverse theory on a firmer basis when we've taken some steps forward in physics that allow us to talk about the very initial instants of the Big Bang, and those theories may tell us whether there were many big bangs or just one, and whether the many big bangs were replicas of each other or whether they were each governed by different sets of numbers.

ah-d

Going back to the Big Bang, there was nothing, and then there was an explosion. Where did all that stuff come from? There was a lot of matter being generated.

mr

Well, we have to understand how we ended up with a large universe from something very small. I don't think we should think of something coming out of nothing. Sometimes cosmologists talk rather loosely about the universe coming from nothing. I think they should watch their language, especially when talking to philosophers, because what the philosopher calls nothing is something straightforward, whereas what the physicist calls nothing is a complex empty space, perhaps, or something that already has laws of nature imprinted in it. So, the physicist's nothing is a lot more complicated than the philosopher's nothing.

The mystery of why there is something rather than nothing is firmly in the province of theologians and philosophers. I'm not holding my breath that they'll resolve it, but it's clear that scientists can never address that question. We can't say what actualizes the equations of physics into a real cosmos. That's not our province at all. But we have to ask, 'Why did something so small end up with this extravagant scale?' It does look like something that is getting a very large amount from a very small investment, as it were.

There's a real sense in which our universe has zero energy. The reason is that there are two sorts of energy in physics. There's Einstein's energy, mc^2, which all material has. But there's a negative energy, due to gravity. For instance, because we're down on the surface of the earth, we have less energy than we have in space. That's called negative potential energy. Many cosmologists believe that the negative energy we possess, due to the gravity not of the earth but of the stars and all the galaxies, may be $-mc^2$. In other words, a negative energy due to gravity may cancel out the mass energy mc^2. If that's the case, the net energy in the universe is zero, and it costs nothing to expand it. And this is the most popular way of understanding how a very small universe can inflate to an enormous size. It doesn't actually require extra energy because the energy in a well-defined sense is always zero, positive mc^2 energy but negative gravitational energy.

ah-d
Are you also saying that all the matter – all the dust and the stars and planets – came out of energy?

mr
Yes. Because physicists believe that different forms of energies convert into each other; that a particle and an anti-particle can annihilate into radiation – and vice versa. So, during the expansion there were all kinds of comprehensive transformations between atoms and radiation and other kinds of energy. The energy is fixed. But its manifestations are changing all the time.

ah-d
One thing that really surprised me in your book was that you said in a cubic metre of space there's only about one-fifth of an atom, but also 400 million photons or bits of light.

mr
That's right. So, our universe is still, in that sense, mainly radiation. That's why we talk about the Hot Big Bang, which is a universe that's dominated by radiation. But actually, one of the mysteries is why it's not pure radiation, because if you were imagining setting up a universe in the simplest way, you might put it into radiation and perhaps equal numbers of particles and anti-particles. But if you did that, then, as the universe cooled down, every particle would annihilate with an anti-particle, producing more radiation, and you'd end up with a universe of pure radiation, not even the one-in-a-billion atoms for every photon. So one of the mysteries

is why the universe possessed this small favouritism for matter over anti-matter, which led to the survival of the matter that we and the stars are made of, even after everything has cooled down. That is one of the things that we are still are slightly perplexed about.

ah-d
Does that sort of mystery have any effect on whether you believe in God?

mr
Different cosmologists react in different ways to that question. I tend to give a rather boring answer. If you'd asked that question of Newton or one of his contemporaries then they would have said something like this: Newton could understand why the planets moved in ellipses, according to his law of gravity, but he couldn't understand why all the planets were going around the sun in the same way and were moving more or less in the same plane. He thought that was Providence. Today we do understand that, because we believe that our sun and solar system started as a spinning, dusty disc, and the dust stuck together to make rocks and then to make planets all going round the same way in a disc.

Indeed, we've pushed the theory far further back even than the formation of our solar system, back to the first second of the Big Bang. But we still, at some stage, have to say, 'Things are as they are because they were as they were.' So, conceptually, we're in the same stage as Newton was, and just as Newton and his contemporaries reacted differently – some were theists, some were atheists – so it is now among modern cosmologists. I don't think modern cosmology has any distinctive contribution to make to our religious perceptions.

The only respect in which being a scientist affects my religious views is that I've become slightly sceptical of any claim to a simple dogmatic answer to any important question, because one thing I have learned is that even a single atom is quite hard for me and for my students to understand. And if it's hard to understand a hydrogen atom, then I'm sceptical of anyone who claims more than a metaphorical understanding of any deep and important aspect of reality. So it makes me perhaps religious in a way – but in a very undogmatic way – impressed with the mystery of the universe, and how little we understand.

I should emphasize that cosmologists are not necessarily being presumptuous when they claim to make statements about our vast universe. That's because what makes things hard to understand isn't how big they are, it's how complicated they are. Inside a star, for instance, everything's so hot that there's no complex chemistry; everything's broken down to its simpler subatomic particles. That's even more true in the early stage of the Big Bang. In contrast, even the smallest living organism possesses layer upon layer of intricate structure. So there's a real sense in which understanding the universe is less challenging than what biologists are trying to do, because biologists are trying to understand far greater complexity, and a star really is simpler than an insect. That's why, incidentally, issues like the origin of life are among the most fascinating. How unique is the life in the universe? Those questions will take some time to answer.

... understanding the universe is less challenging than what biologists are trying to do ... a star really is simpler than an insect.

ah-d

Well, go on then, are we alone? Are there other people out there?

mr

I think the only plausible answer is to say we don't know enough to say whether it's likely or unlikely. I think it's unfortunate that many other scientists adopt firmly polarized opinions when there's no evidence. We must distinguish two questions: first, what's the chance of simple life evolving? And, second, when you've got simple life, what's the chance of that developing into something that we recognize as intelligent? It'd be crucially important if we went to Mars or under the frozen oceans of Jupiter's moon Europa and found evidence of some sort of simple life that originated independently. We'd have to say it's the first stage of life that's simple; but even then, that would leave open the question of whether complicated life was rare or not.

I'm strongly in favour of efforts being made, like those of the SETI Institute in Mountain View, California, to seek evidence for signals that are unambiguously artificial and would reveal intelligent extraterrestrial life, or perhaps artefacts constructed by such life. I wouldn't take bets on likelihood of success. But even if it's very unlikely, it's very important to know whether there is other life out there, and whether there's something apart from human brains that understands mathematics and physics. But we should bear in mind that the only signals we would ever detect would be from intelligence sufficiently like us to be able to understand physics and to understand the world the way we do, and there could be quite different kinds of life that don't reveal themselves in any way. So, a lack of detection does not imply there's no life out there.

This issue affects the way we see the role of the earth in the cosmos. It's traditional to regard the earth as a tiny speck of dust in this vast, hostile cosmos, and us, as humans, being rather trivial. But that might be the wrong connotation, because it could very well be that the odds against intelligent life are so rare that life has not evolved as far as it has here on earth anywhere else in our galaxy.

We know that there are many other planets around other stars. We don't know if there's even simple life on any one of those planets. Even if there is simple life, we don't know whether there's complex life. So, it could be that life is unique to the earth. If that were the case then, I think, it would affect our mindsets in two ways. First, it would make us feel more cosmically important and, second, it would make us feel perhaps even more concerned about the fate of the earth, because the other thing which we learn from astronomy and cosmology is that the universe may have an infinite future ahead of it, and even our earth is less than halfway through its life.

We shouldn't think of ourselves as being the culmination of evolution in any sense. Cosmology does, on this point, offer a distinctive insight that is relevant to religion. Most educated people are aware that we're the outcome of several billion years of Darwinian selection, but there's less awareness of the aeons lying ahead. There's time before the sun dies for as much further evolution as there has been from simple, unicellular life to us. Even if life is now confined to the earth, there'd be time for post-humans to spread through the galaxy and even beyond. Were that the case, then it would be a cosmic disaster, not just a terrestrial disaster, if we were to destroy ourselves. The consequences could resonate through the cosmos, because we'd be destroying not just ourselves but the potentiality of future life.

ah-d

Do you think it matters if human beings disappear?

mr

Well, being one of them, I do think it matters rather a lot. But, more seriously, it matters more if we see ourselves not as a culmination, but as an intermediate stage in evolution. Even if you feel that the human race has many inadequacies, and we'd almost deserve our fate if we didn't preserve our biosphere, then you would still destroy potentialities if you wiped us out.

Imagine the first creature to crawl on to dry land. It might have been a rather unprepossessing and ugly brute. But if that had been clobbered, and no life had ever spread on to the land, you would have destroyed all the potentialities of the life that did develop there. Likewise, if we were to destroy ourselves, then we would be destroying, perhaps, the potentialities of things we can't even yet imagine, and that's the particular crucial thought, if life is, indeed, very rare, as it may well be.

ah-d

You mentioned dark matter. If the entire universe is full of dark matter that we can't see, how can you possibly have missed it? And indeed, how can we see through it to the galaxies on the other side?

mr

When we look at the universe we see shining objects, stars and galaxies, but there's no reason why everything in the universe should shine any more than everything on the earth does. So we shouldn't be too surprised that there may be other things in space apart from the stars and galaxies. We've known for 30 years that there are other kinds of objects in space that don't shine, because when we study the orbits of stars and galaxies around each other, we infer they're feeling the gravitational pull of more stuff than we actually see. This so-called dark matter is important not only for determining the orbits of stars and galaxies around each other, but it's important for the entire universe, because everything in the universe exerts a gravitational pull on everything else. So the more stuff there is in the universe, the more the expansion will be slowing down. For that reason also, astronomers were concerned to pin down how much dark matter there was. And they've succeeded in the last few years, and the results are surprising. It turns out there is indeed dark matter in the universe, but there's not enough to cause the cosmic expansion to eventually come to a halt. The universe will expand forever.

And even more surprising, there's something else in the universe even more exotic than dark matter. There's energy latent in empty space itself, which has a negative pressure to go with it, so it causes the universe to expand at an accelerating rate.

We now know that the universe is made up of about 5 per cent ordinary atoms, about 25 or 30 per cent dark matter, and the rest, that is 65 or 70 per cent, is this dark energy, latent in empty space, which exerts a sort of anti-gravity. So dark matter is important in galaxies, but for the cosmos as a whole, the dominant contribution comes from this exotic energy in space that causes an acceleration. So we have a universe whose ingredients we've pinned down, but they have this rather extraordinary set of proportions that we can't understand at all.

ah-d

You mean that most of the stuff in the universe you can't see? Are you just making it up?

mr

Well, we can't see it and we don't really understand. But we do have fairly good evidence that it must be there because we observe things moving around in a way that indicates the gravity of what we can't directly detect, and we observe the universe to be accelerating and to have the global geometry (that's technically called 'flatness') that indicates the presence of this energy in space. This is a real transformation in the last four or five years.

ah-d
Cosmology's been going for 85 years. Why has it changed so much in the last few years?

mr
Well, it's partly a coincidence that several techniques have come to fruition at the same time, but basically we depend crucially on very precise observations, observing very distant objects with detectors that can detect extremely faint light levels, and also being able to observe from space or from places such as the Antarctic very faint radiation from space. Observations like that have made cosmology a really experimental science, so it's not entirely flaky speculation. These new techniques allow us to observe with great precision some of the fossils of the early universe, which tell us its properties and what it's made of.

ah-d
How is it all going to end?

mr
Most of us suspect that the universe will go on expanding forever, perhaps at a speeding-up rate, but the individual stars and the galaxies will eventually die out, because they will use up all their energy. If you come back and look at the universe many trillions of years hence, you will find it a colder and emptier place, because the external galaxies will have moved far further apart; moreover, most of the stars in our galaxy will have died out, having used up their nuclear fuel, and they will be left as dense remnants of stars.

Some will be what are called white dwarves, some will be neutron stars, and some will be black holes, even more exotic objects. Black holes fascinate astronomers for a number of reasons, and also because they are the most extreme manifestations of gravity. They're places where gravity has overwhelmed all other forces, producing something that has collapsed, cutting itself off from the rest of the universe, but leaving a gravitational imprint frozen in the space that it's left. They also display how counter-intuitive the universe is when we get away from the ordinary, everyday scales.

We're used to the idea of space and time on the scale of the earth, and speeds that we habitually move at, which are never more than a millionth of the speed of light – the speed of a jet airliner. But if you move at close to the speed of light, or could go to a place where gravity is very strong, then space and time behave in a very distorted way. For instance, if you were on the surface of a neutron star, your clock would run 30 per cent slower than a clock a long way away. If you move almost at the speed of light then your clock runs very slowly compared to the clock of someone at rest. And if you go very close to a black hole of a particular kind, then your clock can run as slowly as you like compared to a distant clock. So you could, in what would seem quite a short time to you, observe the entire future of the universe, because your clock is running so slowly.

But you'd have to pick your orbit carefully, because if you were to misjudge and fall into the black hole then you would be cut off from the outside world. Even a light signal that you sent would not be able to escape from the strong gravity, and you would be pulled in right to the centre of the black hole, where there is what is called a singularity. This is where gravity becomes infinitely strong and you'd be torn apart – spaghettified is a technical word for that. To understand what happens at the singularity is going to require an important new step in physics, a step that unifies the quantum theory of the micro-world with Einstein's theory of gravity.

Until we have that unified theory, we won't understand deep inside black holes, nor, incidentally, will we understand the very early instants of the universe, when again we have to worry about quantum effects and gravity together. So black holes are signals that we need some new physics, which we don't yet understand. And also they are objects that really exist in our universe. We can observe a number of them, and as the universe goes on expanding, more and more stars, as they eventually run out of fuel, will end up making black holes. So the universe will become more and more dominated by these cosmic mysteries as it goes on expanding.

ah-d

You've been a cosmologist for most of your life. What most frustrates you about cosmology? What would you really like to solve?

mr

It's actually been exhilarating, because when I was a student in the 1960s, for the first time we were talking about black holes as real objects. We got the first evidence of the Big Bang and it was a good time to start. But the last few years have been more exciting than any previous period, because of the developments that firmed up our picture of the universe, plus also discoveries of planets around stars and other things. I suppose the frustration is that as the frontiers advance, their periphery gets longer. So the more we understand, the more we become aware of new mysteries.

chapter THIRTEEN

Eugenie Scott

Evolution and creationists

Eugenie Scott was born in 1945 in La Crosse, Wisconsin. She studied physical anthropology at the University of Wisconsin-Milwaukee and did postgraduate work in physical anthropology at the University of Missouri, where she first came into contact with creationism. While teaching physical anthropology at the University of Kentucky she led her first successful battle, preventing a Kentucky school board from including creationism in the curriculum. In 1987 she left academia to become Executive Director of the National Center for Science Education, dedicated to "defending the teaching of evolution in public schools" and championing science as a way of knowing. As a humanist, she appreciates the insights science gives us, but recognizes that literature, art and theology also provide ways of understanding the universe. She is a former President of the American Association of Physical Anthropologists, and recipient of the National Science Board's Public Service Award, as well as an Honorary D.Sc. from McGill University.

ah-d

Dr Eugenie Scott is fighting a crusade, trying to keep evolution in education in America. Is that right?

es

Yes, my colleagues and I are trying very hard to keep evolution in the public schools. Evolution is a basic foundational idea of all sciences. People don't think that. They think evolution means man evolved from monkeys. But actually, stars and galaxies evolve, so astronomy is an evolutionary science. And planets evolve – our planet earth has evolved, it has changed through time – so geology is an evolutionary science. And living things have changed through time, which is biological evolution.

Darwin started the trouble with biological evolution. And eventually biological evolution became quite widely accepted. The idea that the earth was very ancient became well accepted. The idea that these hard rocks in the shape of bones and teeth actually represented extinct animals, or former animals, and the idea that life itself has changed through time became accepted too, eventually.

Darwin made two very important points in his book *The Origin of Species* in 1859. One was that living things have shared common ancestors. That was the big idea of evolution, descent with modification – the idea that all living things could trace their ancestry back to, as he put it, one or a few ancestors. So we are all related in a big branching tree of life. His second major point in *The Origin of Species* was that the mechanism bringing about this change was predominantly natural selection.

The idea of natural selection was a truly unique contribution of Darwin's; he put together some ideas that were floating around and made it all make sense. Let's take, for example, the question of why giraffes have long necks. The idea of natural selection is that in a population of giraffes there would be some with longer necks than others, some with shorter necks. If it was advantageous to have a long neck, to reach trees, or roots – foodstuffs that were harder to get if you didn't have a longer neck – then those giraffes with longer necks would tend to live longer and have more offspring than those that didn't. So the tendency to have longer necks would be reproduced throughout the generations. That's one of the real misunderstandings that the public has about evolution. They really don't understand natural selection very well at all. And they don't understand the idea of evolution as being a populational phenomenon.

A couple of years ago in the state of Kansas there was a big dispute about science education standards. The school board tried to take evolution out of science education, which was a big cause célèbre around the United States anyway. There was an old farmer who called in to a radio talk show, and he said, 'I been around animals all my life, and I ain't never seen a chicken turn into a cow.' So he didn't believe in evolution.

But the fact of the matter is that chickens don't turn into anything, and cows don't turn into anything. Individuals are born, they live, they die. It's only groups of organisms, groups of chickens or groups of cows or groups of ferns or whatever that may change through time. It's a difficult concept to grasp.

For natural selection you need three things. You need inheritable, genetically based variation, in a trait like a long neck – some long necks and some short necks. Then you need to inherit, so you need the children, the calves. But you also need a circumstance in the environment that favours one of those variations over others, that favours, for example, longer necks over shorter necks.

ah-d

So if you've got all the apples high in the trees, that favours all the ones with the long necks in that generation?

es

Yes, and that is the selection process.

ah-d

So if you have these three things, variation, inheritance, and selection, then you're bound to get evolution?

es

Well, you are bound to get change, yes. I like to look at evolution as the idea of common ancestry, the idea of a branching tree of life through time. Natural selection is important in producing populations at any given point in time and the variation in those populations. But it doesn't by itself explain this whole big branching tree. The tree is an image that Darwin used, and it's a good one, the idea of all these finer branches coming into thicker branches, coming into a thick trunk, coming into a root, et cetera. But if you think about a tree as a metaphor for evolution, natural selection is out there pruning the twigs and the smaller branches. You can understand the whole structure of the tree only if you look at it in a time frame, because through time you get speciation events, branching events, that produce all of the wonderful tendrils and wonderful populations out at the ends.

ah-d

To get a new species, some trait needs to change, the crocodile needs to grow wings or whatever it is, over a long, long time. New species don't come snap, do they?

es

Speciation is a different part of evolution. It's different from natural selection. Natural selection is a mechanism that provides for adaptation. But to actually get speciation in a population or in a species, you need to have mechanisms arise that produce reproductive isolation, and some of these are mechanisms that might govern, for instance, the time that members of two different populations mate. If one mates in the spring and one mates in the fall, they're not likely to exchange genes and, over a period of time, just because natural selection may be operating differently on them in some fashion or another, they may get to the point where even if they did mate, they wouldn't be able to exchange genes.

So natural selection is involved, but it doesn't create the species. The species itself is a process of these pre-zygotic and post-zygotic isolating mechanisms.

Islands are a wonderful place to look at speciation, because you often have enforced genetic isolation. Islands have been studied since Darwin's time and are still being studied today. The Galapagos, the Canary Islands, there are other places where you can see how an isolated population diverges in time from the mainland population that it evolved from. And actually, evolution can take place more quickly than we thought 30 years ago. Geologically speaking, it's a blink of the eye.

ah-d

Now, tell me about intelligent design.

es

Intelligent design proponents would describe it as a new kind of science that allows you to distinguish between two kinds of things that we see in nature. We see structures in nature that look like they're 'designed'. Structures that get something done, like the many bones of a horse's leg that allow it to run fast, or the many different parts of the mammalian middle ear that allow for the transmission of sound. To have all these parts working together to get something done is what, informally, we mean by design. Some of these design structures can come about through natural process, but the intelligent design people say that some of these structures cannot be explained through natural process, and therefore have to be explained by 'an intelligence'. They're not talking about little green men. The intelligence they're talking about spells his name with three letters, and the first one's a capital G.

ah-d
But are they saying that the ear couldn't have evolved?

es
What they're claiming is that there are certain kinds of design structures that are different from others. Something like the vertebrate eye they would probably concede you could explain through natural selection, but something like the bacterial flagellum, or the structure of DNA, they claim has to be explained by the divine hand of God; they claim that God creates these things, puts all of these parts together so that they work at one time. That's what they would say is the theory of intelligent design – the mathematical and probabilistic and biochemical analyses that allow you to detect one kind of design from another. If you ask me what intelligent design is, I would take a slightly broader view and describe it as a political and religious movement. These folks have a think-tank in Seattle, at a place called the Discovery Center, and the think-tank has a name that I think bears a little reflection, because it helps you understand why many of us in science look at this as a religious movement more than a scientific movement. The name of the centre is the Center for Renewal of Science and Culture.

ah-d
Surely science doesn't need renewing?

es
They feel it does. This is a rather odd kind of focus for a scientific organization. I belong to a lot of scientific organizations, and not one of them is out to renew science or renew culture; this is not what we do in the American Association of Physical Anthropologists, for example. When they talk about renewing culture, what they're talking about is bringing Christian religion back into what they consider an overly secular society. I think fundamentally what they want is to promote theism over materialist philosophy, which is fine, but it's not science. Science is about understanding the natural world.

You can't test supernatural explanation.

But if they were to stand on the corner and say, 'Theism is better than materialism,' they wouldn't get very many people listening to them. If they stand on the corner and say, 'Scientists are lying to you about evolution,' people are going to listen. If they stand on the corner and say, 'Science is anti-God because it insists upon explaining only through natural cause,' people are going to listen. But just to take that latter point, of course science has to restrict itself to explaining only through natural cause. If there's an omnipotent force in the universe, can you come up with some observation that would disprove that? No, you can't. Anything you see in the universe could be the product of an omnipotent force. We don't have a theometer. You can't test supernatural explanation.

When they talk about renewing science, they're talking about going back to the 1700s, they're talking about bringing God in as a causative agent, and still being allowed to call it science. They argue for something they call theistic science. They're trying very hard to package it as a proper scientific and philosophical idea. There have been biologists and philosophers of science who have looked at these arguments and thus far they have not been persuaded. Theologians have also looked at intelligent design and they haven't liked it very much either. Catholics and mainline Protestant Christians accept the idea of what's called theistic evolution, the idea that God created through the process of evolution. This is directly parallel to Newton's idea of gravitation suspending the planets around the sun instead of the angels suspending the planets. God works through natural law is the idea that Newton brought out. And, of course, Darwinian natural selection could be to a religious person another way by which God manages to bring about the universe according to the plan that he has.

ah-d
But then they still have to put God in all the gaps.

es
The god-of-the-gaps argument is that if you don't have a natural explanation for something, God did it. That's already a disaster scientifically, but from a theological standpoint, too, it's objectionable, because if you do find a natural explanation for something for which God is the explanation then that takes God out of the gap; it diminishes God, and leaves God with less to do. So, since about the 1700s, and certainly by the 1800s, there's been a strong tendency in mainstream Christianity to withdraw from the idea of using God as a direct cause of evolution. Obviously, for a Christian, God created the universe. He's in charge of the universe, he sustains the universe, but for most Christians he's not down there tweaking the bacterial flagellum.

I have this image in mind of the Sistine Chapel with God in the clouds and the angels behind him and he's reaching down, and instead of the outstretched hand of Adam, he's got these little bacteria. It's not the kind of image that most Christians would find very attractive. But that's really what the intelligent design people are saying.

This movement looks very much like the creation science movement, with one distinction. The creation science people will tell you that the earth is 10,000 years old, that the Grand Canyon was cut by Noah's Flood. They'll make all these factual claims, and you can test them, and you find out they're dead wrong.

The intelligent design people are very cagey about ever saying what happened. In many respects, they are dealing with the same kind of interventionist God, the same idea that there's something wrong with evolution, as the creation scientists are, but they hide their religious orientation rather more successfully than the creation science people do. I know what the creation science people say about the Cambrian Explosion. I do not know what the intelligent design people say about the Cambrian Explosion, except that they say evolution didn't do it. The intelligent design literature consists largely of arguments against evolution, and especially natural selection, being an explanatory force in nature.

ah-d

They say that DNA is so complicated it has to be created by God. Does that mean that Christian DNA is different from Muslim or Buddhist DNA?

es

No, they would say that DNA was created *de novo*, all at one time, back perhaps at the creation of life itself. When I said before that the intelligent design people are really fuzzy on what happened, they're fuzzy on what happened because they themselves are composed of a real amalgam of individuals.

They are claiming to be scientists and there are a couple of scientists among them. The majority of them seem to be philosophers of science, and there are even a couple of lawyers and journalists. But I think the big tent of intelligent design has been maintained because they are so unspecific about what happened. Some of them deny that any kind of descent with modification takes place. Some of them deny common ancestry. Others accept common ancestry, but don't tend to talk much about what happened. They have some young earth creationists in their ranks who believe that the earth is 10,000 years old. But most intelligent design people won't talk about the age of the earth, because most of the rest of them think that earth is billions of years old, but they want to keep everybody in the big tent.

Some people within the intelligent design movement describe themselves as theistic evolutionists, who believe God created earth through the process of evolution. This is the Catholic position, the mainline Protestant position; it's common in Christian theology. They accept that living things descended with modification from common ancestors, except for things like the bacterial flagellum and the blood-clotting cascade, and these structures which they call 'irreducibly complex', which can't be explained through natural cause, they say. But you end up with a real hodgepodge, because what you have is God reaching down at periodic intervals. God creates the Big Bang, and then God creates DNA and the first replicating molecule. And then that evolves along for a while and then God comes in and creates the bacterial flagellum, and then he comes back in and he creates the Cambrian body plans or something like that.

At least there's a certain amount of coherence to the creation science point of view, which is special creation. God created the earth in six 24-hour days, boom, there you have it.

The intelligent design people don't have a very coherent model, partly because they want to keep the big tent. They would love everyone to link arms against nasty old materialist evolution and then work out their own theological differences later. But those theological differences are enormous, and the result is that they are unable to say what happened. I want to know what did God do and when did he do it, and they won't tell me.

ah-d

Going back to 1802, William Paley said that if you found a rock on the ground it wouldn't be surprising, but if you found a watch, that would be very surprising. You wouldn't expect to find that there by chance.

es

And you would know that because the watch had all these intricate parts that work together, that there had to have been a watchmaker.

ah-d

And then he said, look at the human eye; it is so much more complex; that proves that there must have been a god. Now, isn't this intelligent design precisely that same argument?

es

The intelligent design people are pretty crafty about not mentioning the G word. They just talk about an intelligence, and if you press them on it, they'll say, well, of course we mean God, but it doesn't have to be. But of course it is God. It's a basic design argument, just like Paley said the structural complexity meant God had to have done it, what you find with the modern intelligent design proponents is they're saying that irreducible complexity in a biochemical structure couldn't be produced by natural cause; therefore God did it.

An irreducibly complex structure is one that has a series of parts, all of which have to be present at the same time for the thing to work. It couldn't come about by the incremental process of natural selection; it had to have been created in one go. That is an argument from ignorance, when you think about it. You're saying, I don't understand how the bacterial flagellum could have been brought about by natural cause; therefore it must have been the hand of God. But we shouldn't lose sight of the fact that even an irreducibly complex structure can be produced incrementally by natural selection.

We shouldn't lose sight of the fact that even an irreducibly complex structure can be produced incrementally by natural selection.

To say God did it posits an untestable force. It is simply out of the realm of science to test. If there is an omnipotent force in the universe, anything this force does is compatible with that entity's abilities. So whatever you see in the natural world could be produced by God. This is why we don't consider supernatural intervention in the natural world and still call it science. Maybe God does intervene in the natural world. Maybe God did create the bacterial flagellum. But if we're going to be wearing our scientists' hats, we have to just keep trying to dig out the natural cause.

ah-d

What's happening out there in the schools? Teaching of evolution is not banned in American schools, is it?

es

No. Actually, there have been a number of court cases over the years, including one in 1968 that said, 'Thou shalt not ban evolution'. So it's illegal to ban evolution, but that doesn't mean it has to be taught. There are basically two things going on in the struggle over the teaching of evolution in American public schools; we have two kinds of problem. We have creationism being proposed and promoted, and evolution being repressed. There are efforts constantly being made to get creationism, creation science, intelligent design, some religiously based view, taught as science in the public schools. And there are efforts to get teachers to just skip evolution, or downplay it, or disclaim it by saying it's only a theory, it's not something you have to take seriously. At the National Center for Science Education we try to help teachers and parents and school boards cope with both of these problems.

ah-d

If you take a random public school in small-town America, are they going to be teaching evolution or not?

es

Whenever I talk with British or Continental reporters, they're always astonished at the tremendous amount of decentralization in the American public school system. Decisions about who's going to teach, what we're going to pay them, and what's going to be taught, are made literally at the local school board level, sometimes even at the local school. Sometimes states will provide guidelines for history or science or literature education, but unless the state is contributing a lot to education in a district, those standards usually are advisory, and the local districts can accept them or not. Even if you've got strong science education standards in a district, even if the district says, 'Thou shalt teach evolution', it doesn't mean it's going to be taught. I have found that one of the biggest factors in whether evolution gets taught or not is how secure the teacher feels about teaching it.

If the teacher feels that there's a lot of community hostility toward evolution, that there are lots of letters to the editor of the local paper, for example, about how evolution is a theory in crisis, or how scientists are giving up on evolution, then teachers just quietly start skipping those chapters and not assigning them in the textbooks, and the kids don't get it.

You want me to give you my magic formula for teaching evolution? This is what I think we should be doing. From grades maybe three to six, upper elementary school, we should be teaching kids a little bit about time, so that they get the idea that earth is really quite ancient. There's a whole bunch of wonderful things you can do. You can take a big long string and stretch it across the playground, and mark off the different times, and say, Suzie, you stand over there where the first dinosaurs came and, Margie, you stand over here. You can convert the idea of a long time ago into something that's kinetic, that the kids can understand.

Get across something about time. Get across something about heredity, and get across some kinds of ideas about how traits are passed down from generation to generation. Cats have kittens, kittens look like cats, they don't look like giraffes or dogs, you look more like your parents than you look like anybody else, and that basic heredity idea. Get across something about the idea of adaptation. The moths that are on the dark background are not picked off as easily as the moths against the white background.

Then in junior high you bring out some more information on these three factors: time, adaptation and the idea of heredity. In high school, you intensify these concepts even more. It's very difficult to grasp deep time, it's difficult even for adults to get the idea of billions of years, but keep working at it, and eventually they'll get the idea that there's a lot of time out there. You can also work much more directly on ideas of heredity.

At the high-school level you can get into molecular genetics as well as cellular genetics and really help firm up the idea that the reason why traits get passed on from generation to generation is because of these particulate entities that are passed on in eggs and sperm, and that we understand how this works and are learning more about it all the time. The idea of adaptation is very important too. Now, what happens when you put all of this together – when you have heredity, variation, and adaptation, and you have these populations adjusting to environmental circumstances, and you give this lots and lots and lots of time? Then you have evolution.

But for this, of course, we need a whole lot more coordination among teachers. Teachers have very little academic freedom anyway. If you sign the contract for a school district, you have agreed to teach their curriculum. What we need to do is be sure the curricula that teachers are supposed to teach are good, and that we hold teachers accountable for teaching them.

One idea that we haven't really dealt with yet is the big idea of evolution itself. A lot of people don't really understand evolution. They don't understand the details of natural selection, but often they don't really understand the big idea of evolution itself either, which is that living things shared common ancestors. And that's easy to understand, because we all have ancestors. I don't know my great-great-grandmother, but I can infer that this person lived, because I have a mother, a grandmother and a great-grandmother, and I can infer that there was a great-great there, even though I don't know this person's name. An alternative is that my great-grandmother just suddenly appeared on earth. But it's not unreasonable to infer that I had a great-great-grandmother and a great-great-great … and we can go on and on.

And that's what we do with species as well. We see that we have species that are the product of the passing on of genes through generations, just like I am the product of the passing on of genes from my mother, my grandmother, my great-grandmother and that unknown great-great grandmother. Species are also the product of genes that came down through time, through other species, through parent species, as it were. But we have a lot more cousin species and aunt and uncle species, if you will, because just like your own genealogy is a branching tree, so the tree of life is a branching tree. If we can only get this idea across to people, I think we'll have done a great deal towards helping people understand what the history of life is, and why it's so fascinating.

I think if you're ever going to really understand biology, if you're going to understand why biological phenomena are the way they are, it's by understanding that living things shared common ancestors. Take even something as simple as taxonomy, which everybody learns in sixth or seventh grade – kingdom, phylum, class, order, family, genus and species – that whole nested hierarchy. Why are organisms capable of being organized into species and then genera and then families and then orders? Why is that possible? It's because evolution generates hierarchy. It's because these members of the same genus shared common ancestors further back in time.

The splitting and branching of species through time is what evolution is all about. The reason why we can classify all creatures with warm blood and hair and a single bone in the lower jaw as members of the class mammalia is because the ancestral mammal had a single bone in the lower jaw, and hair and warm blood. And we are the descendant of that, and that's why that complex of three structural features occurs together in living mammals. It is exciting. It's a great shame that kids aren't given a chance to learn it.

ah-d

Does it really matter if it isn't taught in schools?

es

If evolution isn't taught, I don't think students are going to really be able to make sense of biology. Theodosius Dobzhansky, a famous geneticist, once said, 'Nothing in biology makes sense except in the light of evolution.' And by that he meant that biology tells us why things are the way they are instead of some other way. Why is it that every eukaryote or creature with a nucleus, which is almost everything that exists, is based on DNA? Because the first eukaryote had DNA, and we have inherited that from the very earliest ancestors.

The fact that you can describe different groups of organisms as being these sorts of nested hierarchies, which, according to the Linnaean system, we can call families and orders and classes and genera and species, is possible because evolution happened. Dobzhansky said that 'Without the light of evolution, biology becomes a pile of sundry facts.'

ah-d

It's also beautiful, isn't it?

es

Yes. I'm sure that one could live a long and happy life without ever hearing the E word, but one could live a long and happy life without ever hearing Mozart either. I think one's life is a great deal richer for understanding the history of life on earth.

chapter FOURTEEN

Lewis Wolpert

Cloning, counter-intuitive science and depression

Lewis Wolpert was born on 19 October 1929 in Johannesburg. He originally trained as a civil engineer in South Africa, but switched to research in cell biology at King's College London in 1955. His books *The Triumph of the Embryo* (1991) and *Principles of Development* (1998) of which he is the main author, deal with the processes by which the genes in the fertilized egg control cell behaviour in the embryo and so determine its structure. The thesis of *Unnatural Nature of Science* (1992) is that science is not common sense, that common sense is misleading. In *Malignant Sadness: The Anatomy of Depression* (1999) he discusses causes and treatments of depression. He has chaired the Committee for the Public Understanding of Science; he is currently Professor of Biology as Applied to Medicine at University College London, and is researching the mechanisms involved in embryonic development.

ah-d
Lewis Wolpert has been an engineer, is now an embryologist, and writes for the newspapers about practically everything under the sun – an extraordinary all-rounder. Lewis, there's been a lot in the news recently about cloning, ever since Dolly the sheep. And it's not quite clear what one is meant to think about it. Is cloning a threat or a promise?

lw
I think there's been more hysteria about cloning than almost anything else, and it's nonsense, a lot of it. I've been offering a bottle of champagne to anybody who will tell me one new ethical issue that cloning a human being would raise. I also say if you can't think of one, give me two bottles. I'm against cloning because of the dangers of abnormality. You see, in order to get Dolly, they had to do, I think, about 270 trials.

But there's a more serious issue, that from what we know about the way the embryo develops, using the method of cloning to get a child, I think there's a very high risk of things going wrong and having abnormalities, and for the moment I would be totally against it.

ah-d
Why should there be abnormalities?

lw
One of the reasons is that various genes are what are called imprinted, particularly one class of genes that control growth. These particular genes are turned off in the egg but on in the sperm.

Now, when you do cloning, you're taking the genes from somewhere quite different, so this distinction between genes being turned on, being imprinted, or not, is not there. This could have quite profound effects on growth and could lead to abnormalities, which is why I am totally against cloning a human being at the moment.

And I think one has to be very careful that even if the child is born looking normal at an early stage, you can't tell what long-term effects there were, or might be. One really needs many, many more animal experiments over a number of years to see how they continue to develop before I would even consider cloning a human being. No ethics, straight biology.

ah-d

I read somewhere that Dolly the sheep was prematurely aged. Is that right?

I would say that what drove human evolution was technology.

lw

I haven't seen a report on that, but that's one of the possibilities if you take the genetic material that you're going to use for cloning from a cell that's divided a lot. At the tip of the chromosome is a region called the telomere, and there is evidence that as the cells divide, the telomeres get shorter and shorter and shorter. And so this could lead to age-related problems. I haven't heard the story about Dolly. Also, I don't like just sticking with Dolly. You know, one sheep is not good enough for me. I want a lot of Dollys before I take anything that happens with Dolly seriously.

ah-d

You said quite recently in a lecture, and I think in print, that the only difference between we humans and other animals is that we understand causality. This seems a slightly curious assertion.

lw

I wouldn't say it's the only difference but I think it's a fundamental difference. I would say that what drove human evolution was technology.

If we go to the apes, our closest relatives, they use tools, they break nuts and they will use sticks to get ants, but do they understand what they're doing?

There's a very important book by Povinelli, an American, which is called *Folk Physics for Apes*. He's done a lot of experiments with apes to see what they actually understand. One quite nice example is the following.

It's been known for a long time that if you have some boxes and there are some bananas at a height, the apes can learn to put the boxes one on top of the other to get the bananas. However, if you put some stones on the ground below the bananas, they never realize that the boxes are going to topple over. They keep toppling over, but the apes never think of removing the stones. And there are many, many experiments of a much more complex nature.

It's controversial stuff, but I'm totally persuaded by Povinelli that apes do not have a concept of cause. They do not have a concept of force. They do not have the concept that if I take these stones away then the box won't topple over any more. If we come to children, cause is what is known as a developmental primitive. From a very early age, human beings have a concept of cause and effect, so that children will be puzzled if a glass moves on a table and there isn't anything obviously moving it. I think that you cannot make complex tools, that is, have technology of a serious kind, without a concept of cause and effect. And it's in that sense that I think that we fundamentally differ from the apes.

ah-d

That's really intriguing. So you're saying that this essence of causality is imprinted, is hard-wired, in human beings.

lw

It is actually hard-wired. There's quite good evidence that from a very early age, even a few months old, children have a concept of cause and effect.

I don't understand what it means for a child to have this neurological basis, but I think it's the fundamental basis of technology. I also think that technology is what drove human evolution. I know that quite a lot of people say that it was related to social relationships. I think – and there are anthropologists who have argued the same – that language was quite closely related to thinking about causality. If you can't think about tomorrow or yesterday, then it's quite complicated to leave this tool unfinished today and go and work on it tomorrow. And if you don't have a concept of cause and effect, you can't take a stone tip and join it to a spear. For making a fire you've got to have a concept of cause and effect, and that's what apes don't have.

For making a fire you've got to have a concept of cause and effect, and that's what apes don't have.

ah-d

But surely, apes fall out of trees occasionally; they must realize that falling …

lw

No, I don't think so. They don't have a concept of falling, as it were.

ah-d

What do you think about imitation? Do you think that's important in the difference between apes and people?

lw

I'm not an expert on apes. I only read the literature. Some people argue quite strongly that there is culture among apes, but they can neither imitate nor point. I think imitation, both in primates and ourselves, is important, but nothing is as important in my mind as causality. Incidentally, once you have a concept of cause and effect, then you want to explain all sorts of things, you want to explain illness, hurricanes, drought, death – and I think that's the origin of religion.

ah-d

You mean we have to blame something?

lw

I think we find it intolerable to not have an idea as to what was the cause of something that affects our lives. If you go to the doctor when you're ill, it is intolerable for you, or for most people, that he or she hasn't got the foggiest notion what's wrong with you.

And if you get ill, I'll bet you make up a story as to why you get ill. Uncertainty about things that affect our lives is intolerable, so we make up stories about it.

ah-d

We make up stories all the time, don't we? We do something and then afterwards we make up a story. You think it's just an extension of that?

lw

Yes, absolutely. Cause and effect really dominate our lives.

ah-d

So you mean we need religion in order to explain why there is bad weather – or why we die.

lw
I can't say for sure, but I would wish to argue that that's the origin of religion, yes. It's fear that really turns us towards the gods.

ah-d
It seems to me that you're contradicting yourself here, because you've also said that science is not natural, that thinking scientifically is unnatural, and yet you're saying the idea of causality is hard-wired.

lw
Yes, but causality isn't science. Having a concept of cause and effect or knowing what's moving a glass on the table, that's not science. Science is very peculiar. People get quite cross with me for saying this, and say I'm putting them off science.

Only one society ever came across the idea of science, and that was the Greeks. I know that's politically incorrect, but it happens to be historically true. The Chinese did not have science; they had a rather mystical view. The greatest scientist in history, to my mind, is Archimedes. His ideas about floating bodies and levers are simply amazing. The Greeks were the first people to stand back from nature and try to understand the underlying mechanisms. And they were very successful.

I think other people made enormous contributions – Arab societies, all sorts of other societies – but it all originally came from the Greeks. And what I think one has to realize is that unfortunately some Greeks, such as Aristotle, were wrong about many things. Aristotle believed that the world was built on a common-sense basis; therefore he was wrong on everything.

My challenge is this: that anything that comes from your day-to-day experience about the way the world works scientifically is false. The obvious examples are the earth and the sun. Any rational person would say it's the sun that goes round the earth, and yet we now know it's not the case.

Or even quite simple examples show that things are really more complicated. Everyone says it's perfectly reasonable that the tides are related to the moon, because it pulls on the water and it's high tide on the side nearest to the moon. But it's high tide on the far side too!

Galileo struggled with that one. The idea of evolution, that we are here because of random events and selection, I really don't think that fits with common sense. That there are more molecules of water in a glass than glasses of water in all the oceans, I also think doesn't fit.

ah-d
It's quite scary, isn't it, when you put it like that?

lw
Well, I don't mean it to be scary. I think that one of the troubles with people who are not scientists in dealing with scientists is that they expect science to fit with common sense, and it doesn't. Newton's fundamental idea that force causes not motion but acceleration is counter-intuitive. Any reasonable person would say that if something's moving there must be a force moving it.

ah-d
So how do you think about science? What is science?

lw
Oh, I couldn't possibly define science. I'm very bad on definitions. But science is really about finding the underlying mechanisms that determine the behaviour of everything in the world. And as one Nobel Prize winner put it, for any particular set of phenomena, there's only one correct scientific explanation, only one. I think that our problem is to find it. And I think yes, everything is determined by fundamental physics, but you do have different levels of organization. In other words, when I do cell biol-

ogy and developmental biology, I don't have to think about fundamental physics, but I'm not allowed to put something in that contradicts the laws of fundamental physics.

ah-d
You've written recently about depression. Why?

lw
About six years ago I had a severe depression. I'd been low, but I'd never had a severe depression before. This was a depression in which I was suicidal and was hospitalized. There was no particular reason why I should have got depressed at this particular time. Before this, I had taken what I wittily called the sock view of psychiatry, that is, if you're feeling low, you pull up your socks and get on with it.

That depression really was the worst experience in my life; it was terrifying, and I became interested, because it was something I'd never been through before and I couldn't find anything that explained it to me. And actually, one of the newspapers, the *Guardian*, asked me to write about it. I've no idea why. So I wrote an article describing it, and I think I probably had a better response to that article than almost anything else I've ever written. The other thing is, my wife at the time never told anybody that I was depressed. She thought that the stigma would affect my career. I was acutely irritated by this, because why should I be embarrassed about having a very serious illness? That's one of the other reasons – but quite simply, I became interested in depression and wrote a book about it.

ah-d
It's quite odd, though. Most people who are depressed want to hide it.

lw
Well, I am a performer, you know. People say, oh, you're very brave to talk

about your depression. No, not at all. At that particular stage in my career, it didn't make the slightest difference. Also, other depressives were really quite pleased when I talked about it, and I think it's much better if people come out about their depression. It's a serious illness. Even if you do get better, in general. It's a very weird experience and my present bitter joke is, if you can describe it, you haven't had it. You really do enter into a world that bears no relationship to anything else in the day-to-day world.

ah-d

Even if you can't describe it, can you give us some idea of what it felt like?

lw

No, you can't. William Styron's book *Darkness Visible* is probably the best description of depression. And as he says, the word 'depression' is such a mealy-mouthed word. You enter into yourself, you can think about nothing but yourself, you only think about death, there are all sorts of physical symptoms. It's very hard, very, very, very hard to describe. I thought about suicide all the time. My own idea about depression is that it's a normal emotion gone wrong; sadness having become malignant, having got out of control in the same way that normal dividing cells can become cancerous. So in the case of depression, it's a normal emotion, sadness, that's got completely out of control. Whether that's true or not is not clear, but that's my explanation for it.

ah-d

And is that biological, do you think? Have you triggered off some hormones or something?

lw

You can have a purely biological basis for depression. So, for example, if people with various diseases or certain conditions have a high dose of the drug Interferon, there's a high chance they'll get depressed. If your cortisol levels for whatever reasons become very high, you also have a strong base for depression. But there's also the psychological basis; when you're depressed, emotions certainly help to maintain the depression. So it's the biological and the psychological interacting with one another.

ah-d

Do you think you're through now?

lw

No.

ah-d

It'll come back?

lw

I've had one and a half; in fact I'm just coming out of what my psychiatrist claims is another depression.

ah-d

How do you know you're coming out?

lw

Well, the very fact that I'm here being interviewed!

ah-d

That would be impossible, would it?

lw

Oh yes, if you're severely depressed, it would be very difficult. In general, it's mornings that are very difficult to deal with. About 18 months ago, I went down again. I think once you've had one, the probability of having another one is very high.

ah-d

There are masses of drugs given now to control depression. Do they help?

lw

I think they do help. The evidence is quite complicated. First of all, there's no reason to believe that any one anti-depressant is any better than any other. It depends on how the individual responds and on the side-effects. So we all respond rather differently to anti-depressants. The general statement is that if you have a group who are depressed and are given an anti-depressant, about two-thirds will improve, and one-third will not do better. Of those two-thirds that improve, about one-third represents the placebo effect.

The placebo effect is quite complicated. You also have to remember it may not be the placebo because depressions pass anyhow, even if you don't do anything. There's been a lot of fuss recently about anti-depressants causing people to commit suicide. I think the evidence for that is not persuasive. I think anti-depressants really do help.

ah-d

Now, Lewis, I hate to say this, but you're older than I am.

lw

Yes.

ah-d

Quite a lot older. And yet you're still going into the lab every day. Shouldn't you retire and put your feet up and play tennis?

lw

No. First of all, I don't go into the lab. I'm more of a theoretician than any other thing. But I'm still writing. I like being involved with my colleagues, I like being in the department. It gives one a sense of identity. I don't have students or a lab any more, but I have collaborators with whom I do theory. And I'm preparing a book related to what we spoke about a little earlier, about the biological basis of belief. That's what I'm trying to work on. And I think being in a university environment is very attractive. Now I don't want to retire. I'm quite frightened of retiring. It may be the cause of my depression.

ah-d

You love it, don't you? You love the science.

lw

I really do love the science, yes, I do.

Credits

chapter one

Chapter photo – The radio telescope, courtesy of Mullard Space Science Laboratory

Author photo – Professor Jocelyn Bell Burnell, courtesy of Nic Delves-Broughton, IPU, University of Bath

chapter two

Chapter photo – Bathroom Glass, courtesy of the University of Bristol

Author photo – Professor Sir Michael Berry, courtesy of the University of Bristol

chapter three

Chapter photo – Riftia pachyptila – tube worms. Photograph by Colleen Cavanaugh

Author photo – Professor Colleen Cavanaugh, by Jim Harrison

chapter four

Chapter photo – Replication Bomb. Drawing by Lalla Ward

Author photo – Professor Richard Dawkins, by Lalla Ward

chapter five

Chapter photo – Building the blast furnace at Magnitogorsk, 1929. From *The Ghost of the Executed Engineer*, Loren R Graham, Harvard University Press

Author photo – Professor Loren Graham, by Frances Antupit, Cambridge MA, USA

chapter six
Chapter photo – Optical illusions: café wall and duck/rabbit
Author photo – Professor Richard Gregory, courtesy of the University of Bristol

chapter seven
Chapter photo – DNA: photograph courtesy of Photos.com
Author photo – Dr Eric Lander, by John Nikolai: www.official-john-nikolai.com

chapter eight
Chapter photo – Fractals. Image courtesy of Photos.com
Author photo – Professor Lord May of Oxford, by Rob Cousins. Copyright the Royal Institute of Great Britain

chapter nine
Chapter photo – Lady bugs mating: photograph courtesy of Corbis Digital Stock
Author photo – Professor John Maynard Smith, courtesy of the University of Sussex

chapter ten
Chapter photo – Galvactivator: photograph by Webb Chappell
Author photo – Professor Rosalind Picard, by Rick Friedman: www.rickfriedman.com

chapter eleven
Chapter photo – Biological Collection: photograph courtesy of Science Photo Library: www.sciencephoto.com
Author photo – Dr Peter Raven, by Patti Gabriel

chapter twelve
Chapter photo – Time Line for Universe and the Earth, by Jon Lomberg/Science Photo Library: www.sciencephoto.com
Author photo – Professor Sir Martin Rees, courtesy of the University of Newcastle

chapter thirteen
Chapter photo – Detail from the Sistine Chapel ceiling courtesy of Photos.com
Author photo – Dr Eugenie Scott, by Don Melandry, Berkeley CA, USA

chapter fourteen
Chapter photo – Tragedy Mask. Photograph courtesy of PhotoDisk, Inc.
Author photo – Professor Lewis Wolpert, courtesy of University College London

Index